A Golden Decade of Trains: The 1950's

In Color

By Robert R. Malinoski

Photographer & Author

Published by
Morning Sun Books, Inc.
11 Sussex Court
Edison, N.J. 08820
Library of Congress Catalog Card Number: 91-062001
Typesetting by R. J. Yanosey of Morning Sun Books.

First Printing
ISBN 1-878887-06-8

This book is dedicated to my wife Ann who persevered during the trying times of my railroad career. She stood by despite my changeable working hours, those annoying telephone calls in the middle of the night, my having to work some holidays and being called in on days off, etc., etc. She made innumerable sacrifices. Patience has been her virtue.

Acknowledgements

I made many friends since joining the Railroad Enthusiasts in the 1940's. They were slightly older than me and certainly more worldly. During my formative years, they guided me, enlightened me, transported me, and assisted in developing a camera technique. Bob Collins, Don Furler, Dick Loane, Ed May, George Meyer, Bill Newman and the late Dick Prout and others helped me in one way or another. We became lifelong friends.

A towerman is in a profession now almost as extinct as their instrument, the telegraph. I stopped many times at their work places to get a lineup of any train action, to get the ID of unknown trains and occasionally to chat during quiet times. The vast majority were friendly and cooperative.

Locomotive engineers are in an honorable craft. They answered my many questions, were wonderful story-tellers, gave me unauthorized rides on their engines and as Erie Railroad crew mates supported me on many occasions. A great group!

Trainmen rendered invaluable help during my early times on the job, taught me how to do the assigned task and do it the easy way, and especially safely. Passenger men honored my pass beyond its limits and even off-line men "rode" me. They were my co-workers all my career. We fought the battle of seniority and survived.

Gentlemen - I thank you.

Illustrations- by the author, maps sketched by the same.

A Golden Decade of Trains

The 1950's In Color

This book portrays engines and trains of the 1950's, in color, by one person, all over North America. It was perhaps the greatest transitional period in modern railroad history. What was so unusual about the 1950's? Steam locomotives were being phased out in favor of the new diesels. New passenger equipment and even entire new trains were introduced, while freight trains had the benefit of new technical innovations resulting in dramatic speedups. Yes, it was truly a *Golden Decade.*

During the fifties, steam locomotives were still operating in regular revenue service. They ranged from old tea-kettles on the Canadian Pacific, Boston and Maine and even "my Erie," to high drivered, passenger engines to the giant Big Boys of the Union Pacific. The in between types were more numerous, of course. Pacifics were hauling mostly commuter trains and these were usually "plain vanilla" types, those without high maintenance appliances. This type of engine was more economical to operate as they were in service only two to four hours a day and consequently had poor utilization. They were also in the most unprofitable service of all - commutation. Some Hudsons were surviving to cover secondary trains and on standby for peak traffic periods.

Modern freight engines were still running. Big 2-8-4's gave truth to the Nickel Plate Road's claim of *High Speed Service* while that same type hauled coal on the C&O and L&N. Big, trusty 4-8-2's and 4-8-4's operated on several roads and did the usual commendable job. Various types of articulateds served several lines. The B&O, NP and DM&IR worked 2-8-8-4's. N&W utilized mammoth 2-8-8-2's and 2-6-6-4's. Challengers ran on the UP, NP and D&H. The giants of them all, of course, were the Union Pacific's 4-8-8-4's - the Big Boys. Who wouldn't admire those magnificent machines with spinning rods and a stack belching multiple smoke designs and a mellow whistle. It would not be an overstatement that steam locomotives had a universal attraction.

The passenger trains of the 1950's were the glamour girls of railroading. As the decade dawned, several brand new trains had been installed, such as the CALIFORNIA ZEPHYR, PHOEBE SNOW and POWHATAN ARROW. New equipment was purchased for existing trains on the PRR, NYC, UP, ATSF, SP, GN and others. Several new trains were put on including the Santa Fe's SAN FRANCISCO CHIEF and the CP's CANADIAN. Yes, passenger trains were a sheer delight to see, to ride and to photograph.

In the spring of 1950, the major roads of the East began second morning freight service between the New York area and Chicago. That was save-a-day service. Livestock was being hauled in volume and the speedup was extended to that service also. Perishables were crossing the country on expedited schedules as mechanically-cooled cars began replacing the smaller, old fashioned ice-cooled cars. Roller bearings were being applied to an increasing number of cars reducing the hot box problem and radio communication was expanded across systems. Piggyback service began in the middle of the decade and eventually developed into dedicated trains.

Change, progress, efficiency were the key words in the 1950's. Steam engines could not work 24 hours day-after-day like the diesels. They required too much attention in the roundhouses and shops raising their costs per mile of operation. It was a foregone conclusion that diesels would replace the beloved steam engines, and indeed they did. Passenger trains suffered almost a

Robert R. Malinoski after detraining FEC #6, SOUTH WIND, from Miami 11:45 A.M. to Fort Pierce 2:28 P.M. Sunday, Feb. 8, 1953. The 12-car train had PRR observation-lounge car #1129 on the rear end.

similar fate with the introduction of jet planes. POOF went the overnight sleeper business. Freight traffic was gradually declining due to the aggressive trucking industry and within this ten year period the railroads went from the best of times to hard times.

The engines and trains were still extremely interesting. My enthusiasm did not decrease despite the changing times. In fact, I became more and more enamored with the colorful and varied diesel types.

* * * * * * * * * * *
* * * * * * * * * * * *

My fascination with engines and trains began way back when my folks bought me a Lionel set as a Christmas gift. I was also thrilled to see the real trains while living in small towns - Mt. Carmel and-Shamokin, PA. An uncle who worked for the Reading enthralled me with stories about the railroads and I distinctly remember when he showed me the company magazine heralding the new steam-powered, stainless-steel streamlined train in 1937, later christened the CRUSADER. Perhaps my real education began when I acquired my first copy of *RAILROAD MAGAZINE*, the April 1938 issue which described the mile-a-minute runs, the *Trains That Are Making Good* article was about the Seaboard's ORANGE BLOSSOM SPECIAL and also had a roster of the handsome Missouri Pacific engines.

A move to New York City got me started into taking train photos unofficially (to me) meaning I didn't record the data. With my photos of Dec. 9, 1939, I began recording the data, something I continue to this day. A chance meeting in Grand Central Terminal with a member of the *Railroad Enthusiasts* resulted in my joining that group about 1941.

Uncle Sam sent me "greetings" in early 1943 and I was assigned to the Army Air Corps. Basic training was not too hard to take that winter down in warm Miami Beach. I did get acquainted with FEC - SAL power and trains. The next move was to radio school in Sioux Falls, SD and I got to admire those streamlined MILW 4-6-2's on the HIAWATHA. My next stop was at Salt Lake City where I saw UP and Rio Grande activity plus some SP on a trip to Ogden. A short tour at Lincoln, NE got me interested in the Burlington 4-6-4's and 4-8-4's. Soon, flight training was started at Clovis, NM, a great Santa Fe division point. Alamogordo, NM was next with SP activity and I took a ride down to El Paso behind SP 4-8-4 #4449, later to become quite famous since it is still in existence. Salina, Kansas was next where I adored seeing MOP 4-8-2's on the ROYAL GORGE and took short trips to Wichita, Newton and even Kansas City. A short stay at Herington, KS was made more pleasant by spending a few hours at the Rock Island station. All these travels certainly got me more enlightened in the railroad world. We finally settled down overseas in Guam. Our B29 crew, part of a squadron of 15, was one of only five that survived the war.

After World War II, I answered an Erie RR advertisement for temporary work as a yard brakeman. I was thrilled with the "big money" then of $8.54 a day. The pass privilege was a big plus to me and it was used quite extensively.

Informal slide shows in the early 1950's were the "in" thing since 35mm color slides were becoming popular. Many such shows were attended at the *RRE's* Williamson Library meeting room in Grand Central Terminal. My peers projected wonderful material, so impressive that I eventually acquired a 35mm camera at Christmas-time 1952. All of the photos in this book were taken with that Kodak *Retina*.

Over the years cameras and film have improved, many trips were taken, innumerable train slides were added to the collection and a multitude of slide shows were organized. Many times I have been approached with the question of "Bob, when are you going to do a book?" or words to that effect. I always rejected the idea until now. Here it is my "one-man" show." *I hope you enjoy it*

East of Manhattan

The New York Metropolitan area was always one of the greatest railroad centers of the USA. The country's most populous city generated a need for both commuter and long distance passenger service along with a very high freight volume. Good diversity was guaranteed by the dozen large railroads: LI, NH, PRR, NYC, NYO&W, DL&W, ERIE, NYS&W, LV, CNJ, RDG and B&O. East of Manhattan the Long Island operated the busiest passenger service in the entire country. The New York end was electrified with MU's darting back and forth at close intervals. Points farther east such as Oyster Bay, Port Jefferson, Greenport and Montauk were served by steam-powered trains.

(Above) Ten wheeler number 50 was at ease in the Morris Park engine terminal, one of 31 (#20 - 50) of G5s class. This class was the most powerful of its wheel arrangement. Note the MU's in the new grey and aluminum paint scheme. 10:45 AM Oct. 8, 1953

(Opposite page, top) Four of New York City's five boroughs were surrounded by water. Westbound New Haven trains had to come through the South Bronx and cross the East River on the classic Hell Gate Bridge, opened in April, 1917. The bridge was the centerpiece of a $28,000,000 project. NH number 175, THE COLONIAL, from Boston at 8:30 AM was ontime at Sunnyside for its scheduled 1:15 PM arrival in New York's Penn Station. Motor 153 of class EF-3 had 15 cars this trip. 1:05 PM Jan. 12, 1953

(Opposite page, bottom) New Haven Train 182 was running a little tardy from its New York departure time of 10:10 AM. EP3 class 0355 had 10 cars of bulk mail and combine 6015 passing HAROLD tower. Motors 0351-0360 were built by General Electric in 1932 and were nicknamed "flatbottoms." HAROLD tower probably OS'd more scheduled passenger trains than any other tower in the country. The PRR's gigantic Sunnyside passenger train yard is to the right.
11:03 AM Dec. 5, 1953

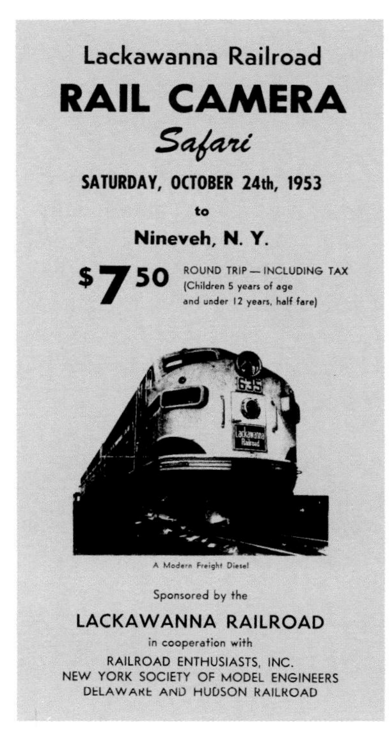

TYDOL
PENNSYLVANIA
355
NEW HAVEN

Delaware, Lackawanna & Western

The Lackawanna at 396 miles was the shortest rail route between New York and Buffalo. It billed itself at one time as the *Route of Scenic Beauty* and finally as the *Route of Phoebe Snow.* Many pleasant rides were completed on their various trains. Hoboken passenger station was the operating nerve center of the east end, with ferry service, the Hudson Tubes, an electrified right-of-way, car shop, export facilities and a major freight yard. MU's ran frequently to mainline Dover and on the Gladstone and Montclair Branches. The non-electrified Boonton Line via Paterson was utilized by a fewer number of commuter trains and all through freights.

(Above) DL&W number 1031, a Monday-through-Friday train, left Hoboken at 2:00 PM on a one hour, twelve minute run to Dover. Pacific 1120, an ALCO graduate of 1920, was temporarily ahead in a race with an MU likewise headed for Dover. Jan. 6, 1953

(Below) DL&W number 11 was named THE SCRANTONIAN and departed Hoboken at 3:25 PM to arrive at its namesake city at 7:15 PM. F3's 802C-B had five head end cars and two coaches in matched attractive maroon and grey paint approaching the GROVE STREET tower. Jan. 10, 1955

Lackawanna hot freight train NE-4, scheduled out of Buffalo, NY at 6:45 PM daily, was a 12 hour overnighter to Port Morris, NJ with an important New England connection via the Lehigh & Hudson River and New Haven railroads. FT's 651-654 (EMD 1945) were originally programmed as pushers out of Scranton. They were later regeared for 45 MPH and put in road service, but were so slow they were called "creepers" by the crews who sarcastically named them after the fastest race horses *Seabiscuit, Citation, Whirlaway and Man O'War*. FT's 654A-B, GP7 965 and RS3 909 were powering 66 cars (including 32 perishables) and caboose 895 getting close to Port Morris, just seven minutes ahead of passenger train number 2, THE POCONO EXPRESS.

12:38 PM Sept. 19, 1959

UNION TERMINAL

The Erie

After three years of Army Air Corps service in World War II, I answered an Erie Railroad advertisement for "temporary" yard brakemen. The job lasted 40 years. No doubt about it, the Erie was "my" railroad.

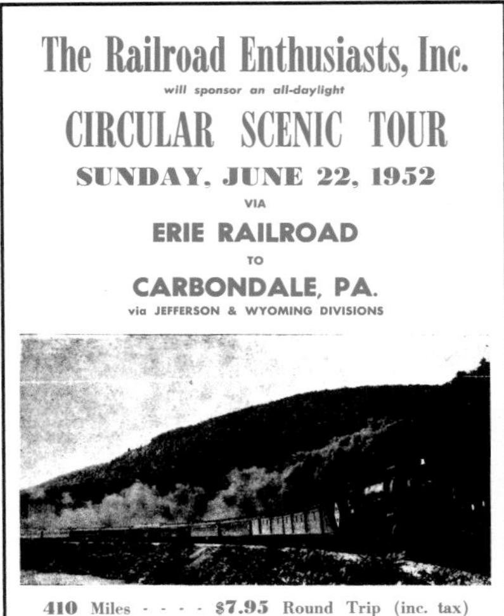

The Railroad Enthusiasts, Inc.
will sponsor an all-daylight
CIRCULAR SCENIC TOUR
SUNDAY, JUNE 22, 1952
VIA
ERIE RAILROAD
TO
CARBONDALE, PA.
via JEFFERSON & WYOMING DIVISIONS

410 Miles - - - - $7.95 Round Trip (inc. tax)
(CHILDREN UNDER 12 YEARS, $4.00)
FOR RESERVATIONS RECEIVED BY JUNE 17, 1952
— $8.25 after this date —

Leaving Jersey City behind a heavy K-5 Pacific locomotive, train will operate via Bergen County Short Cut, thence via Graham Low Grade freight line passing over Moodna Viaduct and through Otisville Tunnel. At Port Jervis we change to diesel power and then travel west along the Delaware Division for the next 100 miles through some of the most beautiful scenery in the East. Our next stop is at famous Starrucca Viaduct built over 100 years ago and still carrying today's heavy trains. At Susquehanna are located the Erie's passenger car shops. Then our train is reversed for the trip over the Jefferson Division to Carbondale, used jointly by the Erie and D&H. Over this line the D&H operate their large 4-6-6-4 type locomotives frequently with 2 or 3 pushers. Midway between Susq. and Carbondale is Ararat, highest point on the Erie, being 2007 ft. above sea level. At Carbondale are located yards and roundhouse of the D&H. From Carbondale the trip is over D&H tracks to Winton where we hit the Jessup Branch of the Erie. In this territory are located many of the large colleries mining anthracite coal. We pass onto the Wyoming Division at Rock Jct. for the run to Lackawanna coming back onto the Main line at that point for the return trip to Jersey City, changing to steam power again at Port Jérvis. This will probably be your last chance to ride behind a steam engine from Jersey City to Port Jervis and to see the largest steam engine in the East in operation on the D&H, as they are to be replaced by diesels later this year.

(Opposite page, top) Monday through Friday was the normal work week for most commuters. However, there was still some Saturday service in 1953. Erie Saturday-only Train 689 left Jersey City at 12:48 PM with a scheduled arrival of 2:07 PM in Spring Valley, NY. Class K1 4-6-2 #2519 with two lightly patronized coaches was starting "up the elevation," a short but nasty grade. US Route 1 to the nearby Holland Tunnel is to the left while the GROVE STREET tower is on the extreme right. The Erie is eight tracks wide here.
Jan. 17, 1953

(Opposite page, bottom) Erie number 607 was a weekday train due out of Jersey City at 5:42 PM taking an hour and 16 minutes to Spring Valley. Number 607 would shortly branch off on to the former New Jersey & New York RR, presently part of New Jersey Transit and named the Pascack Valley Line. The 2521 had four coaches on the drawbridge over the Hackensack River, taken from the HX tower steps. The piers of a previous bridge are visible. Today, high-rise condos replace the open meadows in the background.
5:52 PM Sept. 24, 1953

(Above) The Erie was the first railroad serving New York and Chicago to be 100% dieselized. F7 #711A-B-C-D delivered from EMD in January 1950 helped the cause. Number 711 on "Buffalo 91" symbolled BX-91, with 167 cars and caboose C337 was west of Campbell Hall, NY. BX-91 followed XC-91 and NYO&W BC-1 out of Maybrook Yard. The cars are still passing the "O&W Bridge" in the distance.
1:39 PM Dec. 11, 1953

Pennsylvania Railroad

My first contact with the PRR was in the 1930's as a kid growing up in Shamokin, PA. After seeing three and four I1's struggling east with a heavy train, how could I not admire the Pennsy?

(Above) Most Pennsylvania Railroad freight trains had symbols. This one was an exception, an extra out of the Meadows, NJ yard for Camden, NJ. Its purpose? To get 57 empty reefers to the Campbell Soup plant for loading. L1s 2-8-2 #640 was doing the honors at the photogenic S Curve at Elizabeth, NJ. The clock tower of the CNJ station is visible above the head end cars. 9:59 AM Jan. 22, 1953

(Above) The majority of PRR non-MU eastbound passenger trains that terminated in New York were eventually dispatched through the East River Tunnel to the Sunnyside Yard in Queens Borough for turning and servicing. GG1 number 4905 (GE 1935) had Train 624 in tow where the two tracks spread out to three. This train originated in Lehighton, PA as Lehigh Valley number 24. The train will shortly go underneath those Long Island tracks, also used by the New Haven. 11:13 AM July 15, 1953

(Below) One of the 139 GG1's handled Train 719 out of New York's Penn Station at 12:15 PM to South Amboy, NJ where it was relieved by K4 4-6-2 #920. K4's were noted for their high-mounted headlight, red Keystone number plate centered on the smokebox door, the unique Belpaire firebox and their remarkable performance. The 920 had just passed the "motor stop" sign en route to Bay Head Junction where it was due at 2:21 PM.
1:17 PM April 9, 1956

More PRR

(Opposite page, top) This "busy" photo is an all-time favorite of mine. K4 #1361, later of Horseshoe Curve fame, was wheeling the 10 cars of Train 733, THE BROKER, out of Exchange Place Station approaching Journal Square in downtown Jersey City. Number 733 was passing the twin bores of the Hudson & Manhattan RR called the Hudson Tubes by locals. The H&M later became PATH. NATIONAL JUNCTION tower is in the scene near the MDT banana reefers being interchanged. The Jersey City extension of the NJ Turnpike is still under construction, with New York City in the background.
5:10 PM   May 1, 1956

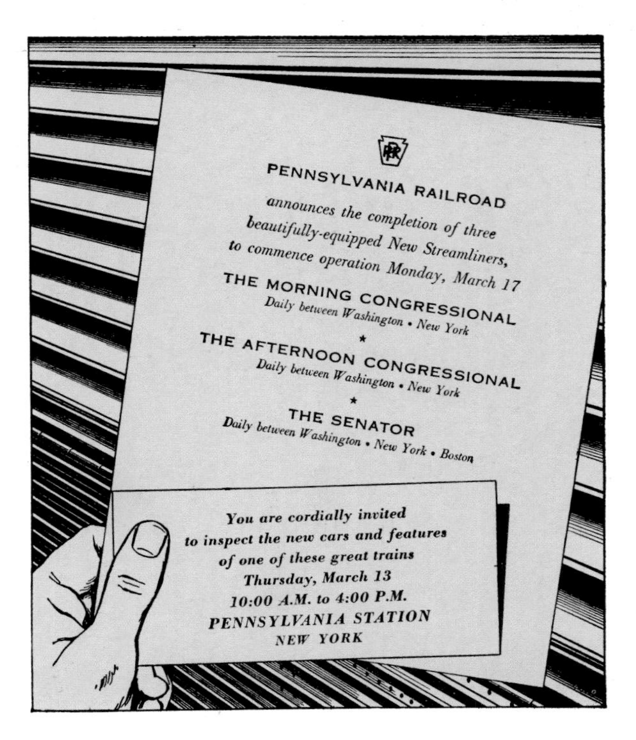

PENNSYLVANIA RAILROAD
announces the completion of three beautifully-equipped New Streamliners, to commence operation Monday, March 17
THE MORNING CONGRESSIONAL
Daily between Washington • New York
★
THE AFTERNOON CONGRESSIONAL
Daily between Washington • New York
★
THE SENATOR
Daily between Washington • New York • Boston

You are cordially invited to inspect the new cars and features of one of these great trains
Thursday, March 13
10:00 A.M. to 4:00 P.M.
PENNSYLVANIA STATION
NEW YORK

(Opposite page, bottom) JC-5 was an empty reefer train pulled by P5A modified 4784. There were 96 cars and cabin car 477552 going west through Iselin on number two track, normally used by eastbounds. Morning Sun Books publisher Bob Yanosey lives nearby.
11:38 AM   Sept. 8, 1955

(Above) Two race trains of two different railroads with two different modes of motive power! It was a pleasant ride on the CNJ's RACE TRAIN SPECIAL from Jersey City to Monmouth Park behind FM Train Master 2409. The adjacent track was occupied by the PRR's PONY EXPRESS with K4 #3747. The shills in the foreground were selling "sure winners." I was fortunate - I lost only ten bucks.
12:50 PM   July 26, 1956

Jersey Central Lines

The CNJ's passenger station in Jersey City was one of the better places to see a good variety of passenger trains. It hosted power from small Camelbacks to large Pacifics hauling some short commuter trains to the heavy long distance versions.

(Above) Steam locomotives with the cabs astride the boiler were called "Mother Hubbards" or the more popular "Camelbacks." They were designed with wide fireboxes to use the slower burning anthracite coal. Many eastern roads rostered them: CNJ, Reading, Erie, D&H, LV, DL&W and a few others. CNJ 4-6-0 #764 was an example, built by Baldwin in 1912 as class L7s, reclassified in 1945 to T-38 (T for ten-wheeler, 38 for tractive effort in thousands). Train 711 departed Jersey City at 2:27 PM for Dunellen.   Jan. 6, 1953

(Below) The CNJ's first 2-8-2's (850-859 class M1s) came from the USRA in 1918. They kept multiplying right up to number 934 in 1925. Number 924 was powering transfer "Run 4" of 28 cars and caboose 91511 across Newark Bay coming off the world's largest four-track drawbridge destined for Elizabethport.
1:28 PM   March 19, 1953

(Above) AJ-2 was an Allentown (5 AM) to Jersey City (10:30 AM) train. "Babyfaced" units 77-74 (Baldwin 1948), lettered for subsidiary Central Railroad of Pennsylvania, had 41 cars and CRP 91539 after setting off 29 cars at E'port.   11:12 AM   Jan. 22, 1953

(Below) My favorite type of steam locomotive was the 4-6-2 Pacific. One of the most handsome was the CNJ series 831-835, Baldwin-built in January 1928, class G3s. They were especially beautiful when painted blue for BLUE COMET service. In my eyes they were an *object d'art*. The 833 with eight coaches and subscription club-car PLAINFIELD made up Train 423 from Jersey City (5:33 PM) to Raritan (6:38 PM) going past Claremont.
5:46 PM   Sept. 9, 1953

5309
BALTIMORE AND OHIO

SCENIC TOUR

of the

Picturesque Catawissa Branch

of the Reading Co.

Railroad Enthusiasts, Inc.
Railway and Locomotive Historical Society
and the
Central Railroad of New Jersey

Invite you and your friends on a trip to
Explore the Scenic Catawissa Branch

SATURDAY, JUNE 20, 1953

A 405 MILE TOUR THRU HISTORIC
AND INTERESTING COUNTRY

$7.95 Round Trip including Federal Tax.
Children 5 years and under 12 half fare

Sixty-five miles of the journey is over the beautiful Catawissa Branch of the Reading Railway, **no longer used for passenger service.**

Opportunity will be provided to inspect and take pictures of steam Locomotives at the C. R. R. of N. J. Round House at Cranford, N. J.

Other stops for picture taking will be made on the Reading Co. at West Milton, Shamokin and Hauks.

DINING CAR SERVICE

Your committee has arranged with the B. & O. R. R. to provide two dining cars to serve Luncheon and Dinner at reasonable prices.

SPECIAL TRAIN WITH OBSERVATION CAR

CNJ continued

(Opposite page, top) A B&O steam engine was kept at the Communipaw engine terminal as protection in case of diesel failure. The call seldom came. Number 5309 drew the unheralded duty this day. The 5309 was one of twenty class P7's (5300-5319) built by Baldwin in April, 1927. Painted an olive green with red and yellow accents, the locomotive was named *President Polk*. It was later rebuilt, reclassified to P7c and painted blue with President name excluded. 1:55 PM March 7, 1953

(Opposite page bottom) The Reading acquired ten two unit 2700 HP sets of EMD FT's in early 1945 (250-259). Reading units were regularly used in pool service with CNJ on the Jersey City - Philadelphia run. Train 685 had a morning departure for the B&O's Eastside Yard in Philadelphia and eventually Potomac Yard in Alexandria, VA. The 255A-B had 32 cars and CNJ caboose 91530 going under the electrified PRR tracks in the Greenville section of Jersey City. 9:10 AM March 21, 1953

(Below) CNJ-RDG Train 619 with all-Reading equipment was named the THE WALL STREET leaving Jersey City at 4:42 PM during the "rush-hour." This train catered to the downtown New York financial district clientele and was due in Philadelphia at 6:25 PM. FP7's 907-906 had the normal consist of two coaches, a diner, another coach, club car *Wall Street* plus CNJ business car number 98. April 9, 1954

NEW YORK CENTRAL
5271

One Day Along The Hudson

The New York Central called itself the *Water Level Route.* Rightly so, as there was hardly a worthy grade on the entire New York - Chicago mainline. The tracks were close to the Hudson River most of the way between Manhattan and Albany.

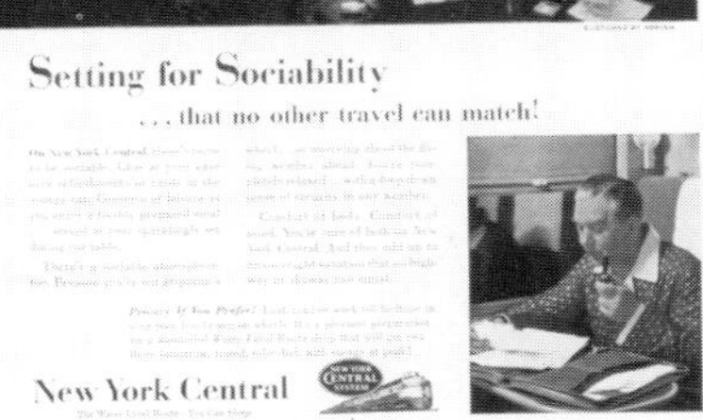

(Opposite page, top) Some sources nominated the NYC 4-6-4's as the most beautiful steam engine ever built. They were christened the Hudson type - the first such wheel arrangement in 1927. The NYC eventually stabled 275 Hudsons, the largest fleet of any railroad. Friday, July 3, 1953 was an extra-heavy traffic day. I rode the first section of number 39 from Grand Central Terminal (12:30 PM) behind motor 260 to Harmon and 4-6-4 #5242 to Beacon. Eastbound number 146 then came in with the 5271. A quick photo was taken before a dash to get aboard for Garrison.

(Opposite page, bottom) THE PACEMAKER also ran in two sections out of New York. This was the second section of number one with the 4402-4401, two of eleven such "Erie-builts" produced in 1949, with 14 cars of all-coach passengers. The passengers on the left side of the train just had a nice view of the West Point Military Academy across the river from Garrison. 5:28 PM July 3, 1953

(Above) Twenty-six passenger trains were seen this day - 14 steam-powered and 12 pulled by diesels. While walking back to the Garrison station to get a train back to New York, number 157 for Poughkeepsie came by with Baldwin 3203 and nine coaches. THE LAURENTIAN, behind ex-streamlined engine 5453, took me to Harmon. It was a good day, all right. 6:27 PM July 3, 1953

Southbound to Miami

During late 1952 I applied for eight trip passes on PRR, RF&P, SAL, FEC, SOU, CRR, L&N and C&O. The chief clerk raised "Holy Hell" about so many. Shortly after they came in, I layed off until further notice in order to use them. The trip was my first long one with a 35mm camera and with only my second roll of Kodachrome. PRR number 101 was boarded at Penn Station with a 12:30 AM departure on February 5, 1953 to Washington and then RF&P number 23 to Richmond, VA. I lingered at the Broad Street Station to see and photograph several trains.

(Above) RF&P number 93 stopped at Hermitage for a power change shortly before high noon. The units were backing to the engine terminal while SAL E8 #3053 and E7's 3032 and 3024 were coupling onto the renumbered SAL Train 3 just as ACL number 26 from Norfolk eased in with N&W 4-8-2 #121, a class J look-alike. For me, excitement reigned supreme to get three different sets of power.
Feb. 5, 1953

(Opposite page, top) RF&P Train 375, THE EVERGLADES, arrived on-time at 1:15 PM from Washington. EMD E8 #1011 and F7B 1159 were painted in attractive blue and light grey with gold lettering and striping. After changing to ACL power, this train continued south carrying me to Petersburg, VA. Feb. 5, 1953

(Opposite page, bottom) N&W's premier passenger train had long been the POCAHONTAS, between Norfolk and Cincinnati. It was an easy assignment for J class #601 with only eight cars on number three out of Petersburg. Those N&W J's at 80,000 lbs. tractive effort, were the most powerful of all 4-8-4's. Later on I met long-time friend and fellow employee Bob Collins who arrived on SAL number 7, THE SUNLAND. We rode number 7 to Jacksonville and then THE SILVER METEOR to Miami. 3:25 PM Feb. 5, 1953

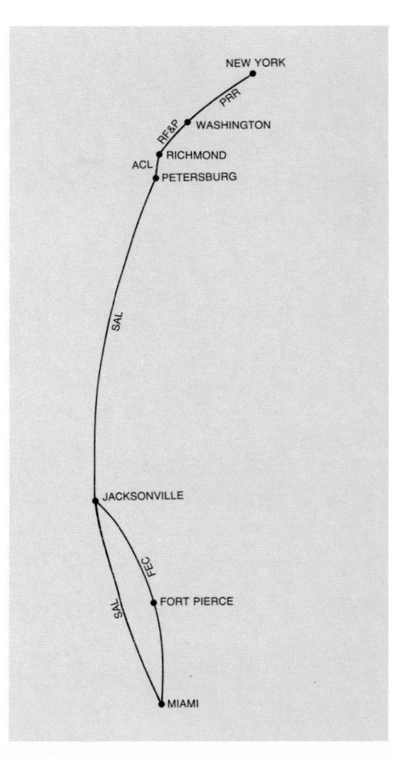

RAILWAY EXPRESS AGENCY
1011

N
82

227
227

In Florida

One reason for going to Florida was to get a peacetime look at Miami Beach where I had my Army Air Corps basic training in early 1943. I couldn't recognize the place ten years later.

(Opposite page, top) Florida East Coast's flamboyantly-painted yard engine #227 just brought THE SOUTH WIND consist into the Miami Station for loading. We rode this train to Fort Pierce for a short stopover. The all Pullman VACATIONER was then our conveyance to Jacksonville.
11:45 AM Feb. 8, 1953

(Opposite page, bottom) The next morning's FEC number 29 took us the five miles to Bowden Yard. Some 0-8-0 switchers were on hand, and the kindly roundhouse foreman even had the 276 spotted on the turntable for our pleasure. Freight train 350 was scheduled in at 1:10 PM and it arrived right on-time with vivid red and yellow F7's 507-552-508 with a light load of only 39 cars, including 14 perishables. Arrangements were made for us to ride engine 271 taking just seven cars to the Seaboard interchange. Feb. 9, 1953

(Above) Jacksonville Terminal RR had several 0-6-0's in colorful blue, orange and silver paint. Engine 14 was contentedly steaming away between station switching chores. Engine 14 was built in the Schenectady plant of ALCO in April 1924, builder's number 65492, as a coal burner now working in mostly oil-burning territory. Pullman car *Shadegap* on the Southern's ROYAL PALM out of Jacksonville at 9:30 PM was our overnight host to Atlanta. 2:20 PM Feb. 9, 1953

A.U.T.
NEW SMYRNA BEACH FLO FEBRUARY 9TH 1953
ALL CONCERNED-
BOWDEN YARD FLORIDA.
NO 350 ENGINE 507-52-08 CALLED 1030 AM WITH 26-0-13

FEC	721	CAB	07			
UTL	97718	ET	SOU	WEST LAKE		
NADX	22090	ER	SAL	AUSTIN		
FEC	606	EB	FEC 07			
PA	439920	EG	SOU			
L&N	49972	SCRAP	SOU	MASSILON		
SP	109179	HHGDS	SAL	BAY RIDGE		
LV	63544	EB	ACL			
AC L	53729	EA	ACL			
CNW	57810	.	SOU			
ACL	55023	.	ACL			
UP	351639	EB	ACL	"351639"		
NYC	147924	.	ACL			
NYC	173684	EB	ACL			
WFE	49039	EB	JAX	JAX SHOPS		
PLE	47952	EG	SOU			
ITC	3107	SCRAP	ACL	SPARROWS POINT		
NRC	794	EXP	NYC	OPEN		
WFE	63647	TOUS	SOU	CHGO	SV VOPD	396
FGE	34815	.	.	JESUP	.	398
BRE	75418	.	ACL	WAYX	.	.
SFRD	32859	.	.	.	.	.
FGE	52004	.	.	.	.	.
.	52588	.	.	.	.	.
.	33462	.	.	.	.	.
.	57741	.	.	.	.	.
GAT	24024	ULS	.	FRANKLIN		
PA	601680	SUGAR	.	PT WENTWORTH		
ATSF	38003	.	.	.		
LN	17678	.	.	.		
CRO	30281	.	.	.		
SOU	25696	.	.	.		
SOU	30540	SUGAR	.	.		
.	25224	.	.	.		
NYC	121449	PITH	.	MONCK		
GATX	23429	MOLS	.	READING		
BRE	74607	CELERY	.	JERS.CITY	R 251 VOPI	K65
FGE	59264	.	.	WAYX	R 240 VOPI	K62
FHI	40743	TOUS	SAL	CHICAGO	SV VOPD	396
FGE	31414	TOUS	.	SAVH	.	398

ELLIS
1212P

The Long Way Home

(Above) The next ride was on the Southern's PIEDMONT LIMITED to Spartanburg, SC. The nearby Hayne Yard had three stored engines: 4-8-2 #1490 and 2-8-0's #394 and 656. The Mountain attracted our attention. She was a big passenger engine, class Ts1, built at the Richmond Works of ALCO in 1919 for the USRA. The apple green and gold paint was fading fast, just like the existence of those engines - three candidates for the cutting torch.
3:30 PM Feb. 10, 1953

Clinchfield Train 37 with year-old FP7 #200 and a consist of two baggage cars and a coach was taken to St. Paul, VA on rainy February 11. Can you believe riding the Clinchfield and not taking a single photo? ASA 10 film, you know. We got a bus to Norton for the night and had two more bus rides the next morning to get to Lynch in order to board L&N Train 22 with FP7 607, a baggage car and a coach. Destination: Corbin, KY. There, we connected with THE SOUTHLAND which had E7 #779 and E8 #794 and 17 cars to Cincinnati, OH.

(Opposite page, top) Friday the 13th was not exactly a sunny day. We admired the very impressive Cincinnati Union Terminal and yard complex completed in Depression year 1933. The fan tracks had a nice array of live engines: C&O 4-6-4 #300; PRR 4-6-2 #5368; B&O 4-6-2 #5212; NYC 4-6-4 #5395 and N&W 4-8-2 #135. Five engines from five different railroads. Not bad! 1:30 PM Feb. 13, 1953

(Opposite page, bottom) We slept the night away in the car *City of Richmond* on C&O number 46, THE SPORTSMAN. E8's 4013-4012 were on the point. The power the next morning was E8's 4008-4009. The train received a good switching at Charlottesville, enabling time to get a shot of our train. Catching the PRR PATRIOT out of Washington at 3 PM, we ended a pretty good trip.
11:20 AM Feb. 14, 1953

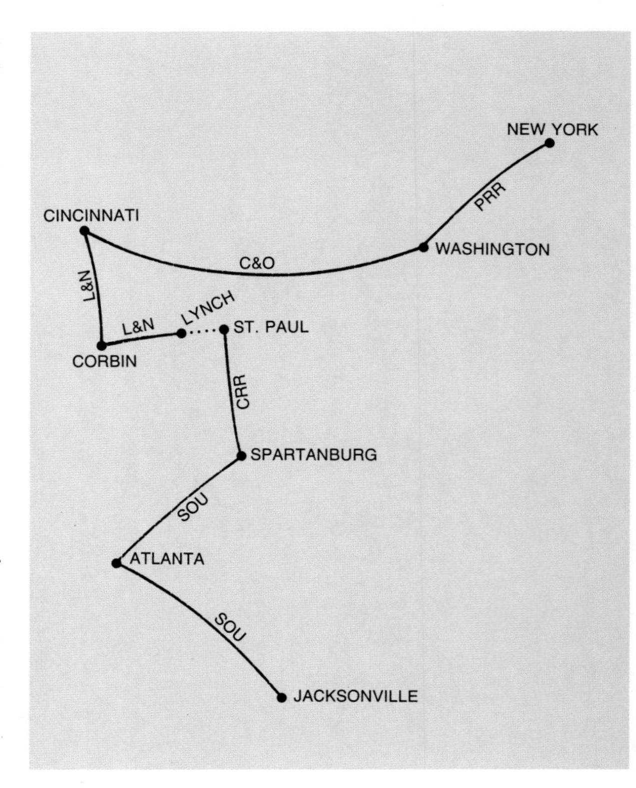

300
300
300

4008
Chesapeake and Ohio 4008

MONON

ERIE

Vacationing by Train

My 1953 vacation started by using the Erie's LAKE CITIES to Jamestown, NY, then a bus trip to Westfield for NYC and NKP action. It was a disappointing stopover, however, account cloudy weather. I endured another bus ride back to Jamestown and then an overnight ride on the westbound ERIE LIMITED to Hammond.

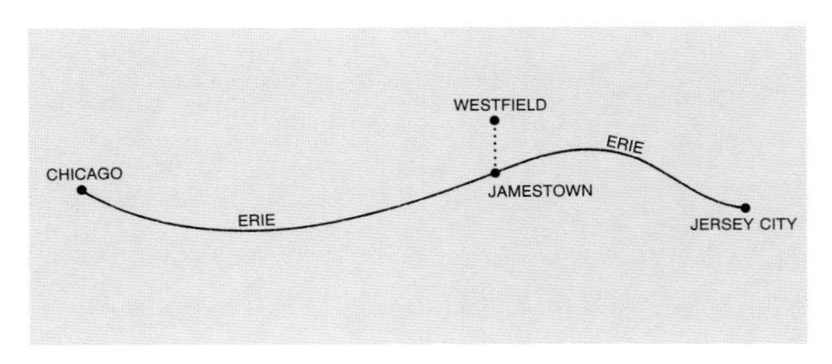

(Opposite page, top) Monon was the popular name of the Chicago, Indianapolis and Louisville RR until the official name change on January 10, 1956. THE TIPPECANOE was tip-toeing through the tulips into Hammond. Number 11 left Chicago at 8:15 AM with F3 #81A - one of only ten such units (81A-B - 85A-B) painted in this attractive red, grey and white. The passenger equipment was converted from US Army hospital cars. 8:52 AM May 30, 1953

(Opposite page, bottom) Erie freight 74 was scheduled to leave Chicago at 8:30 AM and terminated at 3:15 AM the second morning in Croxton Yard, NJ. EMD 1947-built F3 units 802A-B-D were bumped from passenger service upon the arrival of E8's 820-833 in 1951. The train had just passed STATE LINE tower, and Nickel Plate Road tracks diverged to the right. Freight 74's consist of 47 cars and caboose C211 will be reshuffled and enlarged in Hammond Yard with traffic off the IHB. 10:21 AM May 30, 1953

(Above) Doubleheaders were not exactly common on the New York Central's *Water Level Route*. Race horses 5442 and 5418 were starting out of the LaSalle Street Station with the 11 cars and combine rider 200 of fast mail Train 14, due in New York's Grand Central Terminal at 6:30 AM the next morning. 11:25 AM June 1, 1953

MERCHANTS WAREHOUSE COMPANY
WESTERN WAREHOUSING CO.
PENNSYLVANIA RAILROAD
BALTIMORE & OHIO
B&O
NH
468

GRAHAM PAPER CO.
GRAHAM PAPER CO.

"My kind of town, Chicago is. . ."

May 31 was "Valpo" day for me. PRR number 52, THE FORT PITT, took me to Valparaiso, IN where there was plenty of GTW, NKP and PRR action at the interlocking a few miles west of town. GTW number 17, THE INTERCITY LIMITED, with 4-8-4 #6405, took me back to Chicago.

(Opposite page, top) While on Roosevelt Road the next morning, a real surprise of Soo 2-8-0 #468 appeared on a transfer out of Grand Central Station with three baggage cars. Soo steam power was probably the least photographed in color of all Chicago's major railroads. 10:25 AM June 1, 1953

(Opposite page, bottom) The Milwaukee Road's first HIAWATHA train went into revenue service on May 29, 1935. The original power was streamlined 4-4-2's which was gradually upgraded to class F7 4-6-4's. The diesels continued the upgrading with one set each of EMD E6 number 15 and ALCO DL109 number 14 and later on FM's. By the early 1950's three unit EMD FP7 4500 HP sets got the assignment. Number 93C-B-A (EMD July 1950) had the 11 cars of number 101 exiting Union Station. It is almost a sure thing that number 101 will arrive in Minneapolis on the advertised 7:45PM. June 1, 1953

(Above) The Canadian National acquired streamlined 4-8-4's 6400-6404 in 1936. Subsidiary Grand Trunk Western got six (6405-6410) from Lima in 1938 that were essentially the same except for a slight cosmetic change (air scoop smoke deflector). The 6410 was leaving Dearborn Station at 9:40 AM with the MAPLE LEAF, number 20, for Port Huron and Canada's two greatest metropolitan areas, Toronto and Montreal. June 3, 1953

An Illinois Central day

(Opposite page, top) Tuesday, June 2, 1953 was dedicated to the Illinois Central. The first order of business was to get a photo of 4-8-2 #2438 on Train 37 before its 7:15 AM departure. There was a nice background to it with big neon signs of the IC's diamond and another for Armour meats. Secondly, I got on the train for the 75 minute ride to Kankakee. The train became NYC number 414 there, and the new engine was NYC 4-6-4 #5289.

(Opposite page, bottom) A likely photo location was spotted in Monee. I got a Greyhound bus at 9:15 AM for 63 cents to take me there. The fourth train on the scene at 11:11 AM was number 19, THE DAYLIGHT, from Chicago (10:30 AM) to St. Louis (4:00 PM). Units 4024-4020 had the eight car train doing at least 90 MPH. Number 19 was scheduled from Kankakee to Gilman (25.2 miles) in 19 minutes for an average speed of 79.5 MPH. The orange and chocolate color scheme is my all-time favorite.

(Above) Eight of the next nine trains were steam-powered. The exception was freight 72, symboled DISPATCH NC-2, with GP7's 8953-8954, 69 cars and caboose 9817, by at 3:12 PM. At the time this photo was taken to prove that the IC did indeed have diesel freight power. NC-2 was scheduled out of New Orleans at 11:00 PM and was-due at Markham Yard at 2:40 PM the second day. There was no way to get back to Chicago by public transit, so I had to hitch a ride and got one with a guy who could have been charged with DUI. His driving was quite erratic, to say the least, so I requested to get off at the first railroad station served by IC electric trains.

Another Stopover

(Above) My Erie pass was good on CB&Q Train 19, THE COLORADOAN, out of Chicago at 11:00 AM. The power was 9930B and 9911B named the *Silver Mate* with eight head end cars, two chair cars, a diner-parlor and business car *Burlington* carrying the railroad president. At Galesburg I checked my luggage and walked east to the Santa Fe tracks. An ATSF extra west came by with FT's 113-113A-113B handling 127 cars, mostly empty gondolas and box cars, with way car 1863 (just 4 digits then).
3:37 PM June 3, 1953

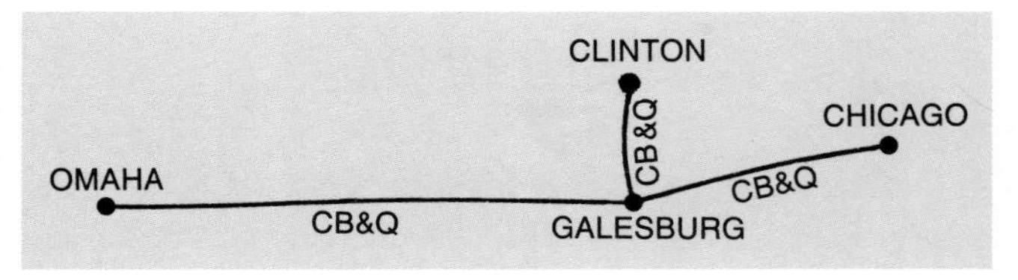

(Opposite page, top) My next move was to the Burlington tracks. There was a decent spot by a ditch when a snake promptly wiggled by me. It really startled me and right in the city limits, too. Train 67A came with FT's 106A-B-C-D (EMD 1944) with 62 loads and 63 empties at 4992 tons and way car 13797. Then it was a real slow ride on CB&Q number 47 out of Galesburg at 6:05 PM to Clinton at 11:15 PM - five hours and 10 minutes for only 93 miles!
3:55 PM June 3, 1953

(Opposite page, bottom) The next day was spent along the C&NW tracks. Six eastbound passenger trains were listed on a lineup. Number 104, THE CITY OF PORTLAND, was right on-time with E7 #5015A, E8 #5022B and 13 streamlined cars passing the tower at 9:00 AM. Two years later the Union Pacific transferred its Omaha-Chicago trains onto the Milwaukee Road. June 4, 1953

AROUND THE WORLD A THOUSAND TIMES!

• Shortly after the first of this year, Burlington's fleet of fourteen diesel-powered, stainless steel Zephyrs attained an unprecedented total in train miles of service—25,000,000!

In rolling up the equivalent of a thousand times around the world, the Zephyrs have proved many things. Their tremendous patronage has shown the public's enthusiastic endorsement of these speedy, luxurious, streamlined trains. Their almost unbelievable record of availability and "on-time" performance, in the face of high speed schedules, has established the might of diesel power. Their beauty and stamina are a tribute to the design and structural strength of these fast trains. (The Pioneer Zephyr, first of the Zephyr fleet and America's first diesel-powered streamline train, has nearly 2,000,000 miles to its credit and has already started on its second ten years of service.)

Yes, America has given the nod of approval to the Zephyrs and the kind of travel they typify. So there will be more of them in the days ahead. They will be even more efficient, more luxurious. For improvement is the mandate of progress.

AN ESSENTIAL LINK IN TRANSCONTINENTAL TRANSPORTATION

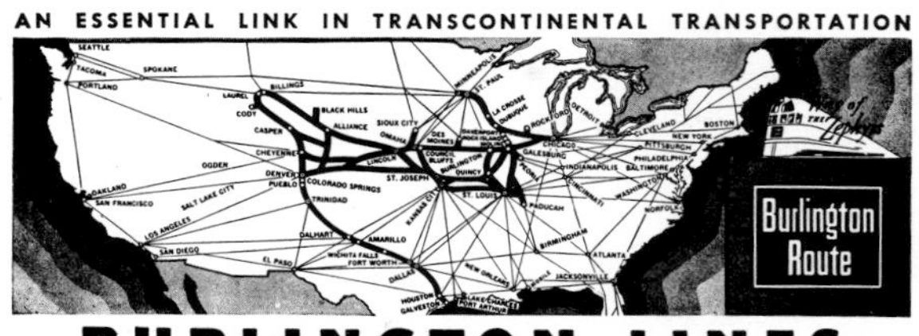

BURLINGTON LINES
Everywhere West

106A
BURLINGTON

Three Railroad Centers

Burlington number 48 was boarded early on the morning of June 5 for the four-hour trip to Galesburg, arriving at 10:30 AM. Three hours later I was back on the COLORADOAN with an 8:35 PM arrival in Omaha.

(Above) Missouri Pacific ALCO PA 8029 with the unique spread eagle on the nose really looked good at the Union Pacific station. The 8029, not quite a year old, (ALCO-GE #78959 July 1952) brought in Train 119 from Kansas City. 8:05 AM June 6, 1953

(Below) I was always fascinated by mail trains. A desire to ride one was accomplished when I rode UP number 5 to Fremont, NE behind three E8's and 18 cars of mail and express, plus a single chair car. It was a fast run as the flagman remarked "the engineer really has his foot on the accelerator." An extra west with rare (to me) three-cylinder 4-12-2 #9035 drifted into Fremont with 74 cars, stopped to take water and departed with that off-beat sound. A few more trains were logged until it was time to get mail train number 6 back to Omaha and then get C&NW's number 202, THE NIGHTINGALE, for St. Paul. 1:37 PM June 6, 1953

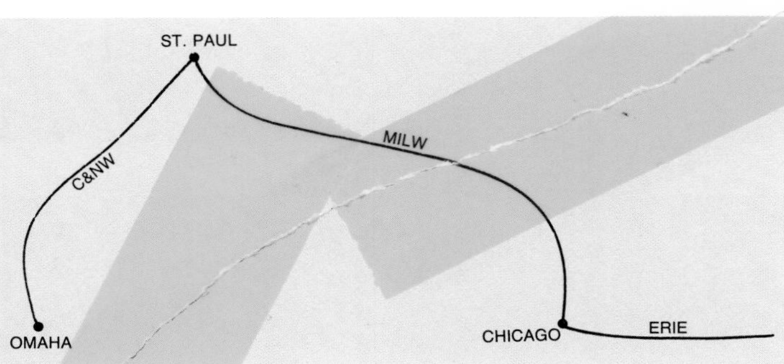

(Above) My hope of getting a photo of Great Northern's FAST MAIL was realized when FP7's #362A-B and F3 #353C took out the 16-car train. That night I rode still another mail train MILW number 56, to Chicago.
9:40 AM June 7, 1953

(Below) A short walk to the CNW station was rewarded with a decent roster shot of rebuilt H1 4-8-4 #3029, a real brute of an engine. The dirty 3029 would shortly take charge of Train 13 for Omaha. I got on Erie number eight that evening for the long trip home. 8:35AM June 8, 1953

303
READING

Hopscotching the East

(Opposite page, top) The Reading bought only six sets of ALCO FA/FB's, 300-305, delivered in June, 1948. The 303A-301B had the 67 cars and caboose 94017 of Jersey City to Hagerstown train second JH-7 on the move through Allentown's EAST PENN Junction. Ten of those units were later traded in on C424's 5201-5210. The tracks to the right lead to the Lehigh Valley interchange.
11:39 AM March 5, 1957

(Opposite page, bottom) In the PRR's freight symbol system, "PG" meant a Pittsburgh Division train. PG-2 started at 43rd St. Yard in Pittsburgh at 11:00 PM, was due to pass SLOPE tower in Altoona at 7:30 AM and terminated in Enola at 1:45 PM. F7's 9548A-B, 9549A had 79 cars and cabin 477324 easing down the world famous Horseshoe Curve. The four tracks have since been reduced to three. 8:26 AM Sept. 11, 1957

(Below) Some sources wondered if the PRSL diesels were ever-painted green. Let this photo be proof positive. Green BLH AS16 #6014, only four months old, was ready to take out Train 1022 at 3:00 PM from Atlantic City, NJ. July 31, 1957

Freights of Distinction

(Below) Most 350-plus mile overnight LCL freight trains were annulled during World War II account heavy military traffic. The New York Central reinstated NB-1 (New York-Buffalo) on July 1, 1946 with the name PACEMAKER. NB-1 was scheduled to make the 429-mile run in under 12 hours. The consist usually was made up of solid vermilion red and grey box cars equipped for 65 MPH service from an original fleet of 425 cars. ALCO FA's 1025-1024 were wheeling 76 box cars, mostly painted in that PACEMAKER scheme, and old caboose 18388 through Westfield, NY. Note the green signal on track 4 for an approaching eastbound local freight.
1:50 PM Oct. 16, 1953

(Opposite page, top) Nickel Plate Road cabooses carried the slogan *High Speed Service.* East St. Louis (Madison Yard) to Buffalo train MB-98 was a good example. Berkshire 734 was barreling into Westfield, NY with 73 cars including six livestock on the head end followed by 46 perishables. 1:37 PM Oct. 16, 1953

(Opposite page, bottom) The Erie's freight service EMD F-unit fleet was numbered 700-713. When the NYO&W ceased operations in March, 1957, F3 units 821A-B, 822A-B were acquired by the Erie, repainted and renumbered into 714A-B-C-D. The 714 was coming out of the Otisville, NY tunnel with train XC-91 (Maybrook 91) with 75 cars Chicago-bound.
11:25 AM Oct. 19, 1958

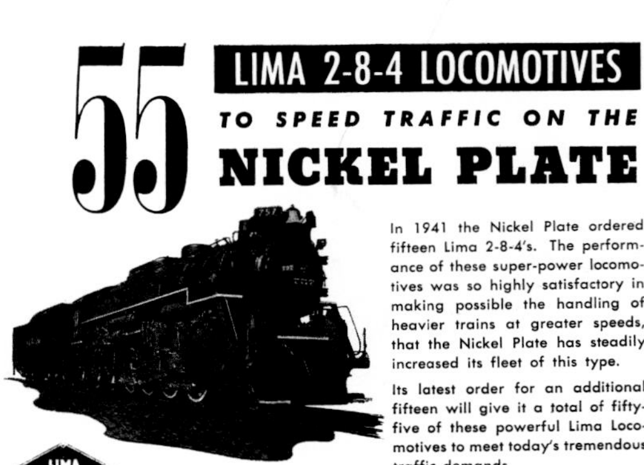

714

READING

READING
READING
READING

In coal country

(Opposite page, top) You look first at the struggling Reading 2-10-2 #3010, but how about that NC&STL candy-striped box car! The train is a "Shamokin Eastern" on a turn to St. Nicholas with eight box cars and 34 coal at the Yellow Hill road crossing west of Mt. Carmel, PA. Engines 3000-3010 were rebuilt into 2-10-2's from 2-8-8-2 Mallets between 1927 and 1930. For a short time the 3010 had Caprotti poppet valve gear.
12:40 PM   Dec. 28, 1952

(Opposite page, bottom) The rear end of the train had the Baldwin-built #3020 shoving hard on caboose 92947. The Reading had a sizable number of hoppers painted with a colorful red, white and black herald pronouncing *America's Largest Anthracite Carrier.*

Reading Lines

(Above) Philadelphia to Newberry train PN-5 had the tonnage limit of 17 empty Cambria & Indiana hoppers and then 33 loads, three more empties and caboose 94058. The trio of F7's 279A-271B-276A was almost at the apex of the 2.6% grade from Gordon up to Locust Summit.
5:20 PM   July 20, 1953

4591
4241
P E

Sensational Sunbury

(Opposite page, top) PRR train S-390 left Northumberland with 2-10-0's 4241 and 4233 headed for the interchange with the Lehigh Valley at Mt. Carmel. However, it had to stop in Sunbury until an Enola-to-Renovo extra passed, powered by I1 #4591 with 109 cars. S-390 then crossed over onto the Shamokin Branch. This "meet" shot is a personal favorite.
9:00 AM April 23, 1955

(Opposite page, bottom) The same S-390 with engineer M. S. Hayes on the 4241 and E. T. Cooper on the 4233 is imitating a volcano going through Paxinos with 85 cars including 54 coal and assigned cabin 477784, which carried the flagman and conductor W. P. Zong. Note the difference in tender lengths. Also note above the box car a little black blurb, which is the diesel of Reading train NT-4. It was almost an ideal two railroad "race" shot.
11:23 AM April 23, 1955

(Above) The PRR mainline through Sunbury splits Third Street. M1 #6761 on Erie-to-Enola freight W-2 emitted a pleasant smoke plume in front of the 81 cars and cabin 477952 on a clear and brisk day.
4:00 PM Feb. 12, 1954

PRR Contrasts in Northumberland County

(Left) BNY-16 was nicknamed the "Beanie" because of its letters. BNY-16 was due out of Buffalo at 3:00 AM on a 25 1/2 hour schedule to Jersey City. FA's 9601A-9615A were rolling along 69 high cars and 477771 through Montandon, just across the West Branch of the Susquehanna River from Lewisburg.
3:27 PM Aug. 20, 1955

"THE ONTARION"

BNY-16

BUFFALO TO HARSIMUS COVE

(Daily)

Buffalo	Lv	3.00 AM	Mon.
Olean	Lv	6.30 AM	Mon.
Renovo	Ar	11.30 AM	Mon.
Renovo	Lv	12.45 PM	Mon.
Rockville "RJ"	Ps	5.45 PM	Mon.
Harrisburg	Ar	6.15 PM	Mon.
Harrisburg	Lv	7.15 PM	Mon.
Morris	Ps	12.45 AM	Tues.
Waverly	Ar	3.30 AM	Tues.
Harsimus Cove	Ar	4.30 AM	Tues.

(Below) Local freight S&L (for former Selinsgrove & Lewistown RR) with H10 #8686 smoking up the scene crossed the North Branch of the Susquehanna leaving Northumberland. Highway 11 bridge over the West Branch of the Susquehanna was in the background. 8:30 AM Aug. 22, 1956

(Above) Motive power rivals join hands! I1 #4628 and EMD F3 #9503A-9502B were leading train S-390's 148 cars (97 loads, 51 empties, 10487 tons) and cabin 477784 with pushers I1 4330 and Baldwin sharks 9706B-9702A by Orphanage. 9:54 AM Sept. 25, 1957

(Below) The Pennsy suffered a motive power crunch in 1956 and 1957 so some power was leased. It was a complete surprise to me to see C&O GP9's #5998-5954-5960 on train S-390. Winter light loading of only 55 cars and ubiquitous cabin 477784 still required the assistance of two more C&O GP9 pushers. 11:40 AM Nov. 9, 1957

M1 Country

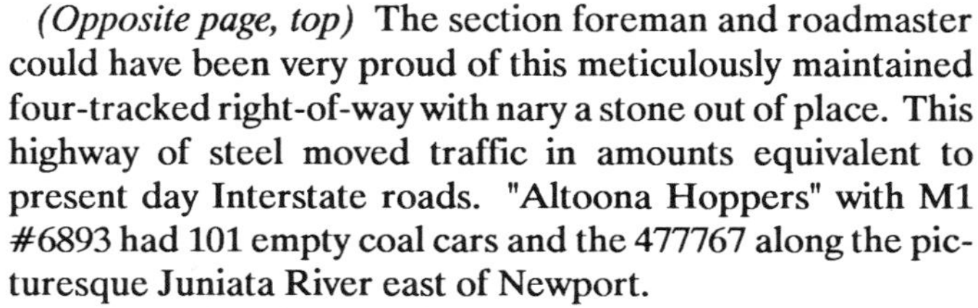

(Opposite page, top) The section foreman and roadmaster could have been very proud of this meticulously maintained four-tracked right-of-way with nary a stone out of place. This highway of steel moved traffic in amounts equivalent to present day Interstate roads. "Altoona Hoppers" with M1 #6893 had 101 empty coal cars and the 477767 along the picturesque Juniata River east of Newport.
9:06 AM May 6, 1955

(Opposite page, bottom) TP-3 with doubleheaded M1's 6762-6928 had a train of empty hoppers going by a caboose hop with 6726 in Marysville, PA, the day after the area was devastated by *Hurricane Diane*.
3:00 PM Aug. 19, 1955

(Above) A going-away shot can be very dramatic and a welcome change of pace. Coal train M-24 from Altoona to Harrisburg had M1 #6979 with 105 cars tied to the drawbar of that coast-to-coast tender coming off the east end of Rockville Bridge. 1:25 PM Nov. 10, 1956

PENNSYLVANIA

Along the Susquehanna River

(Opposite page, top) Some people at slide shows called this a classic photo. It portrays the attitude of the diehard Erie man Bob Collins obviously not impressed with a Pennsy diesel train. BF-4 was not our primary objective - it was the following EC-2 with M1 #6758. Buffalo-to-Enola BF-4 had F3's #9503A-9502B with 21 auto parts box cars ahead of 127 others and cabin 477904 at the west end of Rockville Bridge in Marysville.
4:03 PM May 3, 1956

"Uncle Sam"
VL-2
EAST ST. LOUIS TO ENOLA
(Daily)

East St. Louis	**Lv. (CT).**	2.30 AM	Mon.
Terre Haute	Ar.	7.30 AM	Mon.
Terre Haute	Lv.	8.00 AM	Mon.
Indianapolis (Davis)	**Ar.**	10.15 AM	Mon.
Indianapolis (Thorne)	**Lv.**	1.00 PM	Mon.
Columbus	**Ar. (ET).**	8.15 PM	Mon.
East Columbus	Lv.	10.00 PM	Mon.
Esplen	Ps.	3.10 AM	Tues.
Pitcairn (WG)	**Ar.**	3.50 AM	Tues.
Pitcairn (SZ)	**Lv.**	4.50 AM	Tues.
Altoona (Slope)	**Ar.**	8.35 AM	Tues.
Altoona (Antis)	**Lv.**	9.50 AM	Tues.
Enola	**Ar.**	2.00 PM	Tues.

(Opposite page, bottom) The PRR leased nine Reading class T1 4-8-4's during a 1956 motive power shortage. A Renovo-to-Harrisburg extra was headed by the Reading's #2112 in command of five wide loads ahead of 89 empty hoppers and the 477909 along the scenic Susquehanna River at Millersburg, PA.
12:23 PM Aug. 21, 1956

(Above) VIEW tower in Duncannon, PA was aptly named for it offered one of the best vistas of any tower, anywhere. The Juniata River flows into the Susquehanna within viewing range. Freight VL-2 (for the old Vandalia Line) from East St. Louis to Enola had the 9561A and three others with a high class consist of 12 reefers and 61 mixed freight and the 478165.
8:51 AM Oct. 11, 1956

820
ERIE

Travel by Train

(Opposite page, top) My 1954 vacation trip started on Wednesday, April 21 by riding Erie's LAKE CITIES from Jersey City to Chicago with E8's 827-826 going all the way - the longest diesel run out of the New York metropolitan area. The highlight was a next morning cab ride from Meadville to Kent. Number 5 had a brief stop at Atlantic, PA to let number 8, THE ATLANTIC EXPRESS, with 820-821 and 15 cars to take the siding. Erie's E8's usually had odd numbered units leading westbound, even numbers eastbound.

(Opposite page, bottom) The Pennsy's FORT PITT was used the next day from Chicago on the 58 minute ride to Valparaiso, IN. Grand Trunk first trick operator Jay Blackly was very cooperative by informing me of on coming trains. NKP number 7, THE WESTERNER, with a Buffalo connection off the Lackawanna train of same number and name, had ALCO PA "Bluebirds" 185-187 with four head end cars, two coaches, a diner-lounge, a sleeper and business car number 1 occupied by the NKP president, L.L. White, approaching the GTW crossing.
2:35 PM April 23, 1954

(Above) Second trick operator George Wheeler was just as cooperative. GTW train 495 usually had a diesel ahead of a 4-8-4 to balance power. EMD units F3 #9007 and F7 #9024 were coupled on to the 6313 and the 78 cars with caboose 77147 just after crossing the PRR and approaching the Nickel Plate diamond. GTW number 17, THE INTER CITY LIMITED, with engine 6410 took me back to Chicago. 3:16 PM April 23, 1954

621
THE ROCKET
BOILER WATER
20
18

From the Plains to the Mountains

(Opposite page, top) A ride on Rock Island number 39, THE IMPERIAL, from Chicago to Lawrence, KS was marred account a car developed a hotbox and was set out at West Davenport, resulting in a two hour delay. The lead unit, ALCO DL109 #621 built in October 1941, was re-engined by EMD in 1953 and nicknamed CHRISTINE of Jorgensen sex-change fame. Here she was in Kansas City Union Station.
10:30 AM April 24, 1954

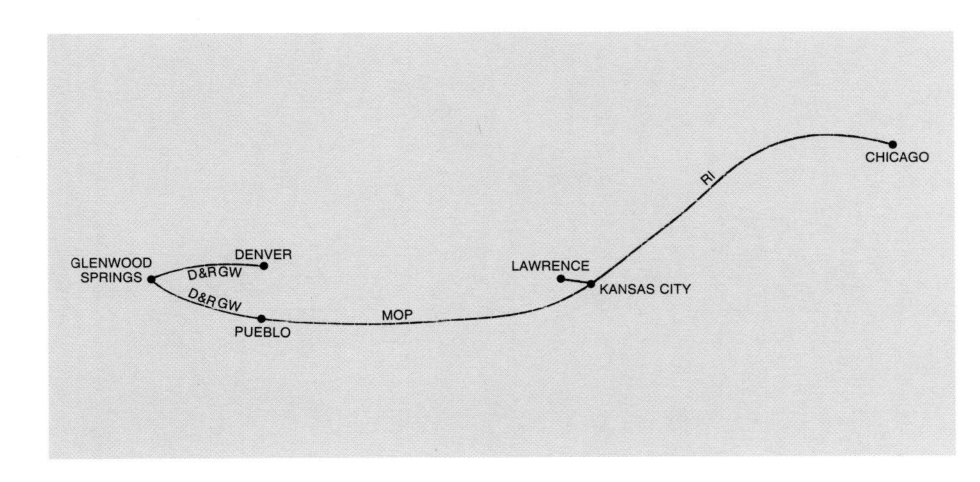

(Opposite page, bottom) Two friends met me at Lawrence where we commenced taking photos of any trains appearing on the scene. One of the best was UP westbound LSS (Live Stock Special) late that afternoon near Lenape, KS with ALCO FA's 1617-1632-1630 hauling 24 livestock cars, 75 mixed freight and caboose 3834. We went back to Kansas City where I got the MOP COLORADO EAGLE for Pueblo.
3:55PM April 24, 1954

(Above) My next ride was the following day on Rio Grande number one, THE ROYAL GORGE, to Glenwood Springs. The next morning's fog was lifting above the red hills when eastbound AD-101 came along with F7's #5681-2-3-4 and eight cars of lambs for North Bergen, NJ on the head end, 52 mixed freight and caboose 01428. A real treat was my ride on the eastbound CALIFORNIA ZEPHYR to Denver behind PA's #6003-2-1.
9:29 AM April 26, 1954

UNION PACIFIC
UNION PACIFIC
U.P.
49311

Ride the

"CITY OF ST. LOUIS"

Domeliner

When arranging a trip between St. Louis or Kansas City and Denver, Salt Lake City, Las Vegas, Los Angeles, or San Francisco (via Ogden), ask to be routed on this beautiful Domeliner . . . the only through train between St. Louis and Southern California.

You travel in relaxed comfort. If you have need for a car at destination, Hertz or Avis Rent-a-Car service is quite inexpensive.

Sherman Hill

(Opposite page, top) Union Pacific number 57 powered by 4-8-2 #7012 took me from Denver to Cheyenne, WY. It was shocking to see GP9's and F3's coupled ahead of Big Boys on westbound trains. That procedure was almost insulting to those wonderful 4000's. It was another surprise to see an eastbound lumber train drift into Cheyenne with the unusual combination of F3's 1450-1464C-1428C in front of UP's first 4-8-4, the 800. Twenty-one stock cars containing lambs were ahead of 80 mixed freight and caboose 3944. 5:27 PM April 27, 1954

(Opposite page, bottom) THE CITY OF ST. LOUIS originated at 4 PM in its namesake city. The train traveled on the Wabash to Kansas City and then the UP to Los Angeles. The next morning found it cresting Sherman Hill and exiting Hermosa Tunnel with PA's 605-604 and FM Erie-built #655 trailing a 14-car consist. 12:35 PM April 28, 1954

(Above) The UP's $16 million Dale Cutoff project, when opened in 1953, reduced the westbound 1.55% ruling grade to .82%. A lineup revealed that three westbound trains with 4000's were coming. One of the *Kings of Steam,* #4003, had the 115 cars of a drag on one of the cuts and fills that dominated the west end of said Dale Cutoff. 5:40 PM April 28, 1954

6513
NORTHERN PACIFIC
NORTHERN PACIFIC

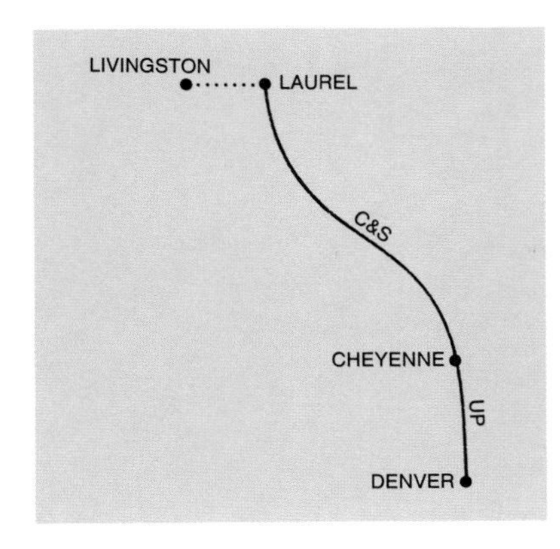

Snow - Steam - Semaphores

(Opposite page, top) It was a 1:50 AM Cheyenne departure on C&S number 29 that took me north to Laurel, MT on April 29 and then on a wild bus ride in a snowstorm to Livingston. A car was rented and I headed east to Elton which resulted in getting Northern Pacific freight "Second 603" with 4-6-6-4 #5143 splitting the semaphores amidst a snowy landscape. Four cars of livestock were on the head end ahead of 53 other cars totaling 3865 tons.

9:38 AM   May 1, 1954

(Above) NP freight 611 had a clear block getting close to Livingston. EMD F7's 6016A-B-C-D had one of the nicest primarily black paint schemes. The train was under the charge of conductor Warren McGee, a noted rail photographer, in caboose 1322 behind the 85 loads and six empties weighing 5120 tons.

5:33 PM   May 1, 1954

(Opposite page, bottom) In 1947 NP's NORTH COAST LIMITED got streamlined Raymond Loewy-designed equipment in an attractive two-tone green paint scheme. Livingston was an engine change point. Number 25's new set of power, 6513C-B-A, had a 10 car load up Bozeman Pass and points west. That night I took the MAINSTREETER to Glendive.

1:35 PM   May 2, 1954

Mainstreeter
NORTHERN PACIFIC
Mainstreeter
NORTHERN PACIFIC

Badlands to Chicagoland

(Opposite page, top) After a 24-hour stopover in Glendive, it was "all-aboard" NP's MAINSTREETER again. The 10-minute stop at Jamestown, ND enabled time to shoot Pullman eight-section solarium-lounge car *Garrison Club*. A transfer was made at Staples, MN to number 56 with motor car B16 and coach A161 to Duluth at 12:35 AM. 4:15 PM May 4, 1954

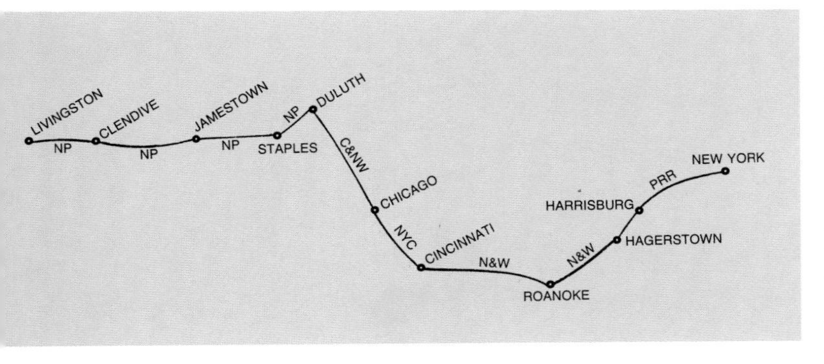

(Opposite page, bottom) Another rental car took me up the hill to Proctor to see some DM&IR power. "Rest Point" was an ideal place to record the frequent activity showing Duluth in the background. DM&IR "up" train with ex-B&LE 2-10-4 #702 was crawling up the hill with 38 empty ore jennies and hack C168. The CHICAGO LIMITED of the C&NW was my transportation to Chicago.
5:54 PM May 5, 1954

(Above) A clean steam engine on the C&NW in 1954 was an oddity. Class E2 4-6-2 #2912 was not only clean but remindful of an Erie K5A with those Boxpok drivers. Train 528 from Janesville had a new paint scheme cafe-lounge car plus nine coaches entering North Western Station. My homeward journey was via NYC number 406 to Cincinnati, N&W Pocahontas to Christianburg, local 24 to Roanoke and number 2 via Hagerstown and PRR to New York. 9:40 AM May 6, 1954

679
683
BALTIMORE AND OHIO

B&O
BALTIMORE & OHIO
BALTIMORE & OHIO
B&O
638701

Interlude at Sand Patch

The late and great David P. Morgan, editor of *TRAINS*, and renowned photographer Phil Hastings took a steam-searching trip to the Midwest in September, 1954. Their chronicles were the inspiration for my 1955 trip, the first long one by car - my 1953 Ford.

(Opposite page, top) The first stop was a short one around Duncannon, PA. The second one was at Manila on the B&O where the budding trees displayed sure signs of the oncoming spring. An ore train with brand-new Geeps 679-683-694 was lensed lugging 70 hoppers of ore from Cuba. It would soon enter the 4475-foot Sand Patch Tunnel.
3:12 PM May 6, 1955

(Opposite page, bottom) The rear end was just as interesting as the head end, possibly more so to steam fans. Big Sixes 6152 and 6145 emitted heavy smoke plumes to prove they were shoving hard on caboose C2439. The area is still inundated with those cinders. That third track has been removed - it was formerly used by westbound drag freights like this one.

(Above) For some reason I forgot, no photo was taken of the head end of freight 396 at Sand Patch, 2258-feet above sea level. However, the rear end was more attractive. Baldwin shark-nosed 857 and 4-8-2 #5584 were serving as pushers against caboose C2440. 4:00 PM May 6, 1955

765
765
765

Steam Steam Steam

(Above) Ohio was a great state in which to see steam power in action in 1955. Just about every major road still dispatched steam engines there, from 0-8-0's to 2-8-8-4's. SIAM, also called ATTICA Junction, was the site where the B&O east-west mainline crossed the north-south PRR Columbus-Sandusky Branch. A PRR extra north with C&O-design 2-10-4 #6469 had 116 cars of coal in tow eventually to be dumped into lake boats.
8:57 AM May 8, 1955

(Opposite page, top) Engine 765 was contentedly steaming away the early morning hours at Bellevue, OH posing in the preferred position of rods down after being washed. The Nickel Plate Road had 80 of these fine 2-8-4's, numbered 700-779. This engine was assigned to take out freight NC-3 later in the day. Engine 765 continues to this day in fan trip service.
8:05 AM May 9, 1955

(Opposite page, bottom) B&O class P7D 4-6-2 #5304 was one of only two known engines to be streamlined TWICE. NYC 5344 was the other one. The 5304 was streamlined for the ROYAL BLUE in 1937, de-streamlined in 1940 and re-streamlined in 1947 for the CINCINNATIAN. Her assignment this day was the eight-car number 57 to Louisville, here charging out of North Vernon, IN.
9:32 AM May 17, 1955

550
C&IM
550
6284
C&IM

701
701
C&IM
701

Central Illinois

(Opposite page, top) The weather deteriorated, to put it mildly, over the next several days and was not at all conducive to ASA 10 film. May 14 dawned crystal clear in Springfield, IL. The C&IM dispatched sparkling clean class F4 2-8-2 number 550 with a rider coach to Peoria to pick up a 10-car train of Boy Scouts arriving on the Rock Island. The railroad was completely dieselized by year end. Sister engine 551 was donated to the St. Louis National Museum of Transport. 7:50 AM May 14, 1955

Form 48G002

TELEGRAM

OMIT EVERY UNNECESSARY WORD

SPGFLD MAY 13-55.
RJW
RFC

C&E EXTRA 550 NORTH

YDS LIST ENG CREW CONDR HAIGH AND ONE BKMAN FOR 630 AM SATURDAY MAY 14TH GIVE THEM ENG 550 AND CONVOY CAR 401TO GO DIRECT TO PEORIA AND TURN ENG AT BRIDGE JCT AS INSTRUCTED BY P P U WHO WILL HANDLE COACH 401 PUTTING SAME ON REAR OF 11 CAR PSGR TRAIN AT UNION STATION WHICH EXTRA 550 WILL GET AND LEAVE UNION STATION AT 1020 AM FOR SHOPS. FIGURE WILL NOT BE NECESSARY TO TAKE ENG TO HOUSE UNLESS OTHERWISE ADVISED JOINT SHOPS HAVANA PEKIN AND PEORIA.

E PT 915 AM

(Opposite page, bottom) The C&IM was chiefly a coal hauling road, with most of it moving north to Havana to be barged further on to owner Commonwealth Edison. One of only four (#700-703) 2-10-2's in class H1, the 701 passed a colorful caboose in olive drab and red. The 701 later posed for a fine roster shot - see *RAILFAN & RAILROAD* of June and July, 1990 for further information.
4:30 PM May 14, 1955

(Above) Burlington's hefty class M4A 2-10-4 #6324 was built by Baldwin in May, 1929. It posed in the ideal rods-down position and low, late afternoon sun in Centralia. This engine, while at Beardstown, was featured in the Morgan-Hastings pieces *Smoke Over the Prairies* in July, 1955 *TRAINS*. 5:40 PM May 16, 1955

820
L&N
LOUISVILLE & NASHVILLE
LOUISVILLE & NASHVILLE

Trip Finale

(Above) The L&N, like the PRR, had a famous class M1 steam engine. The L&N acquired fourteen 2-8-4's from Baldwin in 1942 (#1950-1963) at about $192,000 apiece. Baldwin delivered an additional six M1's in 1944 (#1964-1969). The final order of 22 went to Lima in 1949 (#1970-1991) with an escalated unit price of around $268,000. These last were the most modern of all since being equipped with most major appliances. The 1988 and 1966 were at DeCoursey Yard, KY prior to taking out "Time Freight 45."
3:20 PM   May 17, 1955

(Opposite page, top) Time freight 45 cleared Beray, KY about two hours later. An unexpected northbound train suddenly appeared with F7's 820-711 and RS3's 128-143 with 115 loads of coal and caboose 927.   5:25 PM   May 17, 1955

(Opposite page, bottom) The 2-8-4 type was usually named a Berkshire but there were exceptions. The C&O, which preferred the name Kanawha for the river and county in its territory, was one. Class K4 #2741, husky but dirty, out of Paintsville, KY erupted smoke like Mt. St. Helens as it passed through Thealka with 121 coal and caboose 90259 tipping the scales at 9987 tons tied behind the tank.
10:06 AM   May 18, 1955

Some pertinent data:

WE ARE NOW delivering an order of 22 locomotives of the 2-8-4 type to the Louisville and Nashville Railroad Co.

Modern steam locomotives like these will show a good return on their investment. With planned scheduling, they can deliver more ton-miles of freight per dollar of investment than any other type of locomotive.

Class M-1
Service: Freight
Road Nos.: 1970-1991
Tractive Force, with Booster: 79,290 lb.
Cylinders: 25" x 32"
Drivers, Diameter: 69"
Weight on Drivers: 267,500 lb.
Weight on Front Truck: 52,900 lb.
Weight on Trailing Truck: 127,700 lb.
Total Weight of Engine: 448,100 lb.
Fuel: Soft Coal
Grate Area: 90 sq. ft.
Steam Pressure: 265 lb.
Tender Capacity: 22,000 gal., 25 tons

DIVISIONS: *Lima, Ohio* — Lima Locomotive Works Division; Lima Shovel and Crane Division. *Hamilton, Ohio* — Hooven, Owens, Rentschler Co., Niles Tool Works Co. *Middletown, Ohio* —The United Welding Co.

PRINCIPAL PRODUCTS: Locomotives; Cranes and shovels; Niles heavy machine tools; Hamilton diesel and steam engines; Hamilton heavy metal stamping presses; Hamilton-Kruse automatic can-making machinery; Special heavy machinery; Heavy iron castings; Weldments.

B-2073

West Shore in Bergen County

(Opposite page, top) The NYC's old West Shore line between Weehawken, NJ and Albany, NY was long operated as "The River Division." Most eastbound freights had the symbol "WS." The fall foliage was fading away fast as WS-4 appeared on the scene. It had FA's 1069-1070 and 103 cars rolling down the four-track mainline in Teaneck, NJ. The line is now single-tracked with a siding. 11:30 AM Oct. 28, 1954

(Opposite page, bottom) The New York Central leased some U.S. Army diesels for several months in early 1956. USA units B-2073, B-2047 and 1819 have the 100 cars and bay window caboose 20349 of train WV-1 just past FY tower and on the seldom-opened drawbridge crossing Overpeck Creek. This is a short double-track section sandwiched between four-track territory. 2:55 PM March 15, 1956

(Above) The West Shore operated a way freight between Weehawken and Dumont called the "Ping Pong" account bouncing back and forth across the mainline to serve the various sidings. The four tracks here in Ridgefield Park dwindle to two. The NYS&W double track is on the left. Lightning-striped BLH 6234 had only 10-cars heading east after a March snowstorm. 2:30 PM March 21, 1956

838
BALTIMORE AND OHIO

Hopscotching Around

(Opposite page, top) The most important point on the PRSL was Camden, NJ. The engine terminal was a busy place - witness the multitude of power. There were many engines lettered PENNSYLVANIA including four 0-6-0's, two 2-8-0's, and three 4-6-2's while the READING was represented with seven G3's and a solitary G2SA.
10:30 AM Aug. 27, 1954

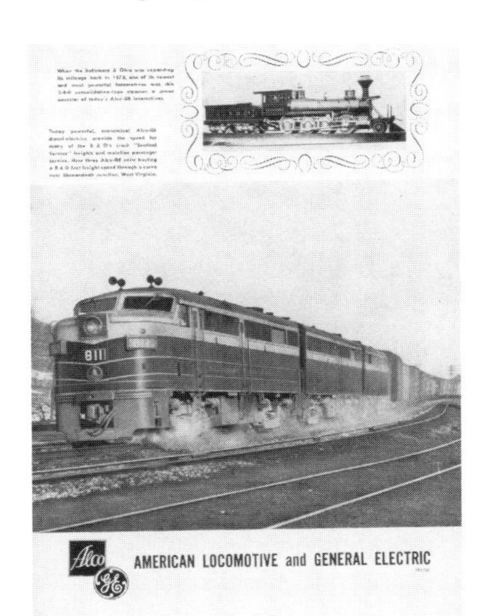

(Opposite page, bottom) Some circles considered the B&O's blue, black and grey with gold striping paint scheme as the nicest of any eastern road. Who are we to object? ALCO FA's 838-823X thought they could make it up the Sand Patch grade at Manila with a heavy 55-car drag and the C2453 assisted by 2-10-2 pushers 6113 and 6222. This train was struggling to stay ahead of three *Timesavers* - the first only seven minutes behind.
9:48 AM March 3, 1956

(Above) This Sunday morning, Feb. 12, 1956, was pretty much a day off for both Reading people and engines at Gordon, PA. The Reading's extraordinary T1 class 4-8-4's numbered 2107, 2112, 2114, 2119 and 2103 were at ease outside the square enginehouse. The T1's were built with boilers from 2000-series 2-8-0's.

5654
BALTIMORE AND OHIO

SANTA FE
5035

DETROIT — PONTIAC — DURAND — CHICAGO — DURAND — MUSKEGON

READ DOWN / READ UP

①79 ♥ Ex. Sun.	①77 ♥ Ex. Sat & Sun.	57 Daily	19 Ex. Sun.	21 Daily	Miles	TABLE 82 Eastern Time	22 Ex. Sun.	122 Sun.	54 Ex. Sun.	56 Daily	①72 ♥ Ex. Sat & Sun.	①74 ♥ Ex. Sun.
P.M.	P.M.	P.M.	P.M.	P.M.			A.M.	A.M.	A.M.	P.M.	A.M.	A.M.
6.15	5.20	11.15	4.55	1.40	0.0	Lv Detroit, Mich. (Brush St.) . Ar	7.35	7.35	9.30	5.45	8.10	8.40
6.23	5.30	11.25	5.04	1.49	4.1	Milwaukee Jct.	7.24	7.21	9.19	5.36	7.59	8.29
6.27	5.36	..	5.08	f 1.53	6.6	Highland Park	7.17	b 7.13	f 9.11	f 5.31	7.51	8.24
6.32	5.45	..	5.16	..	10.9	Ferndale	7.08	..	f 9.05	b 5.26	7.43	8.14
6.34	5.48	..	5.18	..	11.9	Pleasant Ridge (10 M. Rd.)	7.04	..	..	..	7.39	8.12
6.36	5.51	11.41	5.25	2.05	13.2	Royal Oak	7.01	7.05	9.02	5.23	7.36	8.09
6.38	5.55	..	5.28	..	14.3	Oakwood Blvd. (12 M. Rd.)	6.57	..	f 8.52	b 5.19	7.31	8.05
6.45	6.02	11.50	5.40	2.11	17.7	Birmingham	6.51	6.58	8.48	5.15	7.25	8.00
6.49	6.07	..	5.44	..	19.9	Charing Cross	f 6.46	..	f 8.40	b 5.09	f 7.20	7.55
6.51	6.10	..	5.46	..	21.2	Bloomfield Hills	f 6.43	..	f 8.37	b 5.07	f 7.18	7.53
7.00	6.20	12.02	5.55	2.25	26.3	Ar Pontiac . Lv	6.35	6.45	8.30	4.58	7.10	7.45
P.M.	P.M.	12.20	6.00	..		Lv Pontiac . Ar	6.15	6.34	8.07	4.53	A.M.	A.M.
		..	f 6.07	..	31.1	Drayton Plains	..	..	f 8.00	..		
..	..	..	f 6.10	..	33.3	Waterford	..	..	7.56	..	..	..
..	..	..	f 6.12	..	34.3	Windiate	..	..	..	..	..	..
..	..	..	f 6.14	..	35.2	Clarkston	..	..	7.52	..	..	..
..	..	..	f 6.22	..	41.4	Davisburg	..	..	7.43	..	..	..
..	..	12.43	6.28	2.48	46.4	Holly	f 5.52	f 6.11	7.35	4.31	..	..
..	..	f 12.50	6.35	2.55	50.7	Fenton	5.46	f 6.05	7.26	4.25	..	..
..	..	..	f 6.41	..	55.3	Linden	..	..	7.18	..	..	..
..	..	..	f 6.50	..	62.6	Gaines	..	..	7.08	..	..	..
..	..	1.10	7.00	3.20	66.9	Ar Durand, Ⓜ (81) . Lv	5.25	5.45	7.00	4.05	..	..
		A.M.	P.M.	P.M.			A.M.	A.M.	A.M.	P.M.		
		5 Daily		17 Daily			6 Daily		14 Daily	20 Daily		
		A.M.		P.M.		(Table 81)	A.M.		A.M.	P.M.		
..	..	1.40	..	3.42	0.0	Lv Durand, Mich. Ⓜ E.T. . Ar	5.15	..	2.08	3.52	..	..
..	..	2.10	..	4.14	32.4	Ar Lansing " Lv	4.36	..	1.34	3.20	..	..
..	..	2.51	..	4.44	50.9	Charlotte "	4.03	..	d 1.03	2.52	..	..
..	..	3.20	..	5.10	76.7	Ar Battle Creek Ⓜ " Lv	3.30	..	12.35	2.23	..	..
..	..	3.05	..	4.48	101.1	Vicksburg . C.T.	1.52	..	d 10.58	12.50	..	..
..	..	..	..	5.07	117.5	Marcellus	..	..	..	f 12.32	..	..
..	..	3.33	..	5.27	130.4	Cassopolis "	1.20	..	c 10.31	12.17	..	..
..	..	4.06	..	5.50	153.4	South Bend "	12.48	..	10.05	11.52	..	..
..	..	5.04	..	6.40	197.8	Valparaiso, Ind. "	12.02	..	9.19	11.07	..	..
..	..	6.02	..	7.26	243.3	Chicago Lawn, Ill. "	11.12	..	8.31	10.20	..	..
..	..	6.30	..	7.45	253.6	Ar Chicago, Ill. C.T. . Lv	10.50	..	8.10	10.00	..	..
		A.M.		P.M.		Dearborn Stn.	P.M.		P.M.	A.M.		
		57 Ex. Sun.					22 Ex. Sun.					
		A.M.					A.M.					
..	..	2.25	..	..	66.9	Lv Durand, Mich. Ⓜ E.T. . Ar	3.40	..	..	..	..	..
..	..	2.50	..	..	78.5	Owosso	3.05	..	..	..	..	..
..	..	..	..	..	88.8	Ovid	..	..	..	..	..	..
..	..	3.16	..	..	98.1	St. Johns	2.41	..	..	..	..	..
..	..	..	..	..	112.8	Pewamo	..	..	..	..	..	..
..	..	3.45	..	..	124.2	Ionia	2.08	..	..	..	..	..
..	..	..	..	..	132.5	Saranac	..	..	..	..	..	..
..	..	..	..	..	139.4	Lowell	..	..	..	..	..	..
..	..	..	..	..	148.0	Ada	..	..	..	..	..	..
..	..	4.35	..	..	157.5	Ar Grand Rapids Lv	1.20	..	..	..	..	..
..	..	5.00	..	..		Lv Grand Rapids Ar	12.55	..	..	..	..	..
..	..	6.10	..		194.3	Ar Muskegon Hgts. / Muskegon, Mich. E.T. Lv	12.01	..	..		..	..
		A.M.					A.M.					

TABLE 83 MUSKEGON — MILWAUKEE — FERRY SERVICE

(No Motor Vehicles Carried)

READ DOWN — Daily — READ UP

P.M.	P.M.	A.M.	A.M.		P.M.	P.M.	A.M.	A.M.
11.00	4.00	11.00	5.00	Lv Muskegon, Mich. . E.T. . Ar	3.30	9.30	3.00	6.30
4.30	9.30	4.30	10.30	Ar Milwaukee, Wis. . C.T. . Lv	8.00	2.00	7.30	11.00
A.M.	P.M.	P.M.	A.M.		A.M.	P.M.	P.M.	P.M.

NOTE: Tickets sold only on board ship and subject to U.S. tax

1956 Trip

(Opposite page, top) The Boston and Maine gave Baldwin four orders for a total of 18 very handsome 4-8-2's numbered 4100-4117. When diesels replaced steam they were literally put on the auction block. Power-short B&O bought the first 13 and gave them minimal cosmetic changes and new numbers 5650-5662. The T4 class 5654 was at New Castle Junction, PA on May 26, 1956. She was built in 1935 as B&M 4104 and later named the "BEE AND EMMA."

(Opposite page, bottom) The Pennsy leased twelve Santa Fe 2-10-4's in 1956 for coal train service from Columbus to the dumper at Sandusky. An extra south of 50 empty hoppers was powered by ATSF 5035 (Baldwin 1944) and PRR 6491 (Altoona 1943), comparatively modern locomotives. It is obvious that the Santa Fe engine is larger than the Pennsy J1. A talk to the friendly tower operator resulted in him relaying by radio my wish for the lead engine to make smoke. They delivered!

12:38 PM May 28, 1956

(Below) The Grand Trunk Western had three passenger trains due at Durand, MI at about the same time. Muskegon-to-Detroit number 56, due 4:20 PM to 5:05 PM, had 4-6-2 #5627. Number 21 in the reverse direction was scheduled from 4:30 to 5:00 PM and was powered by the 5633. Mainline number 20, THE MAPLE LEAF, was arriving ten minutes late with streamlined 4-8-4 #6410. It was not a normal scene.

5:05PM May 29, 1956

6038
6039
THE ROUTE OF THE
SuperChief
PACIFIC FRUIT EXPRESS
PFE

3132

A Study of 4-8-2's

(Opposite page, top) Some engines, like women, were more beautiful than others. GTW's class Ule 4-8-2's were real beauties, with a beetle-browed feedwater heater atop the smokebox, all Boxpok drivers and a Vanderbilt tender. Number 6038 teamed up with 2-8-2 #3752 to pump up the air prior to departing east from Battle Creek with a 90-car extra.
8:50 AM May 30, 1956

(Opposite page, bottom) Not many 4-8-2's were built after the New York Central's last order of class L4b 3125-3149. Lima-built in 1943, they were the most modern of the type, enabling them to perform dual-purpose duty. Mohawk 3132 was in Hoosier territory at Stockwell, IN with an extra east under the 70-car limit, thereby negating the use of a "full-crew" member. 4:40 PM May 31, 1956

(Above) The Illinois Central's 2400-series 4-8-2's with 73 1/2-inch drivers were built primarily for passenger service but also worked fast freights. Diesels eliminated their presence on passenger trains. The 2422 with an auxiliary water tank was passing the train order office at Foster, IL, just south of the large Bluford Yard. Train CN-1 consists of two cars of Hormel meat head out of the 97-cars. 6:00 PM June 2, 1956

Steam in Kentucky

(Opposite page, top) No coal was mined on Sundays so there was usually nothing to move out of Ravenna on Mondays. LWC was one term used by railroaders to denote an engine running "light with caboose." "Big Emma" number 1958 is running light with old wooden caboose 639 northbound on the Red River Bridge near Sloan, KY. 3:22 PM June 4, 1956

(Opposite page, bottom) Sister Big Emma number 1953 was doing the job for which she was designed - lugging 131 cars of coal out of Eastern Kentucky to DeCoursey Yard, near Cincinnati, OH. Solid-black F3's 2550B-2501A running backwards were pushing on the rear end.
3:53 PM June 5, 1956

(Above) There wasn't a diesel to be seen at Paducah at this particular time. It was an all-steam panorama of the IC's engine terminal. The most obvious engine was the freshly-painted 4-8-2 #2527. 11:35 AM June 3, 1956

NOW . . . *More Famous Than Ever!*
L&N's Country Ham Breakfast . . .

These hams are carefully selected and cured especially for the L&N Railroad on a farm at Pewee Valley, Kentucky, where they are properly aged to assure that old-fashioned "salt-cured" flavor. Then, expertly prepared according to our own recipe by experienced chefs, served with real red gravy, hominy grits, fresh eggs cooked to your taste, and choice of hot bread, this is a breakfast you won't forget!

Equally Famous is the L&N SEAFOOD PLATTER — Available on our New Orleans trains, this rare treat includes soup du jour, soft shell crab, oysters, Jumbo shrimp, fried trout or broiled mackerel, and other delicacies...fresh from Gulf Coast waters. Try it!

For delicious food and fine service, travel via L & N

LOUISVILLE & NASHVILLE R.R.

2770

1644
FORD
COST

Vacation Finale

(Opposite page, top) C&O ran an eastbound set off and pick up train out of Russell, KY called the "Bulldog." It was entering the yard at Handley, WV with 2-8-4 #2770, 52 cars and caboose 90140. A yard crew with an 0-8-0 had a cut of cars to switch and fortunately just cleared the scene.
11:40 AM June 6, 1956

(Opposite page, bottom) The 2 PM lineup at Handley, WV on June 6,1956 read: "1644 called 12:30 . . . 112-3 out Hinton . . . picks up 10 loads at Gauley." The C&O had the country's only 2-6-6-6's, named the Allegheny type, class H8 built by Lima in 1941 with 67-inch drivers, 110,210 pounds of tractive effort and a total weight of 751,830 pounds. One hundred and twenty-five cars were tied to the 1644's tank passing through Montogomery, WV. It may be an extra but it could be called a "New River Train."
5:20 PM June 6, 1956

(Below) The NYC experimented with low-slung, efficient passenger *Train X.* They named it *Xplorer,* put it in service between Cleveland, Columbus and Cincinnati and hoped for the best. It never happened. Here it was running as Train 422, Cleveland-bound, passing through the PRR-operated interlocking at Worthington, OH. 5:54 PM June 7, 1956

New England

(Below) A 2-6-0 Mogul in service in 1954 was a rare bird, indeed. B&M 1427 of class B15b was ready to leave Boston's North Station with train number 2109 for Saugus and Lynn with a combine and three coaches. Those intruding diesels soon sent the 1427 to the deadline forever. 5:13 PM Sept. 3, 1954

(Opposite page, top) A return train ride from Montreal was on daylight CN-CV number 336, THE AMBASSADOR. CN 4-6-2 of class J7a was the power to St. Albans, VT where I dashed off to get this photo. The distinctive station and four-tracked, 350-foot train shed was built in 1869. The new power was CV ALCO RS3 #1860.
11:55AM Sept. 18, 1955

(Opposite page, bottom) Bob Collins and I went on a trip that would test one's stamina. We left NJ at 11 PM on Wednesday, July 18, 1956, drove all night and in the morning came across this CV way freight with 2-8-0 #465 doing some switching at Swanton, VT. The clean-as-a-whistle class N5a was sporting the white flags of an extra. We got only one other train later on. It a was tiring drive back to New Jersey and Collins had to be at work at 11 PM. Fortunately for me, it was my day off so I got to bed.
8:00 AM July 19, 1956

CENTRAL
VERMONT
5265

Fall, 1956

(Opposite page, top) Bob Collins and I started another trip to the West on Erie number 5, THE LAKE CITIES, to Chicago. It was then another overnight ride on Milwaukee's number 19, THE ARROW, to Council Bluffs, IA. We were met by friends who took us in a hurry to the Union Pacific yard in time to get an action shot of the train we rode. Number 19 had the 99A and FM Erie-built 7B with three head end cars, a coach, our Pullman sleeper *Oak Post* and two deadhead sleepers.
9:49 AM Sept. 30, 1956

(Above) We boarded UP number 27 at Omaha that evening for a third consecutive overnight trip; this time to North Platte, NE. We spent the morning there and got a UP eastbound leaving town with my favorite 4-8-4 class FEF-1 doing the honors. Engine 815's assignment that day was to move 101 manifest cars and caboose 3997 to Grand Island and FAST. 12:10 PM Oct. 1, 1956

(Opposite page, bottom) Five of us then headed west in Art Stensvad's Buick. We chased a westbound drag of 124 empty PFE's with 4-6-6-4 number 3958, built by ALCO in 1942. A westbound local freight with 2-8-2 number 2262 sprouting an oversized stack was in the hole at Ogalalla, NE for the drag, setting up a dandy meet shot.
1:10 PM Oct. 1, 1956

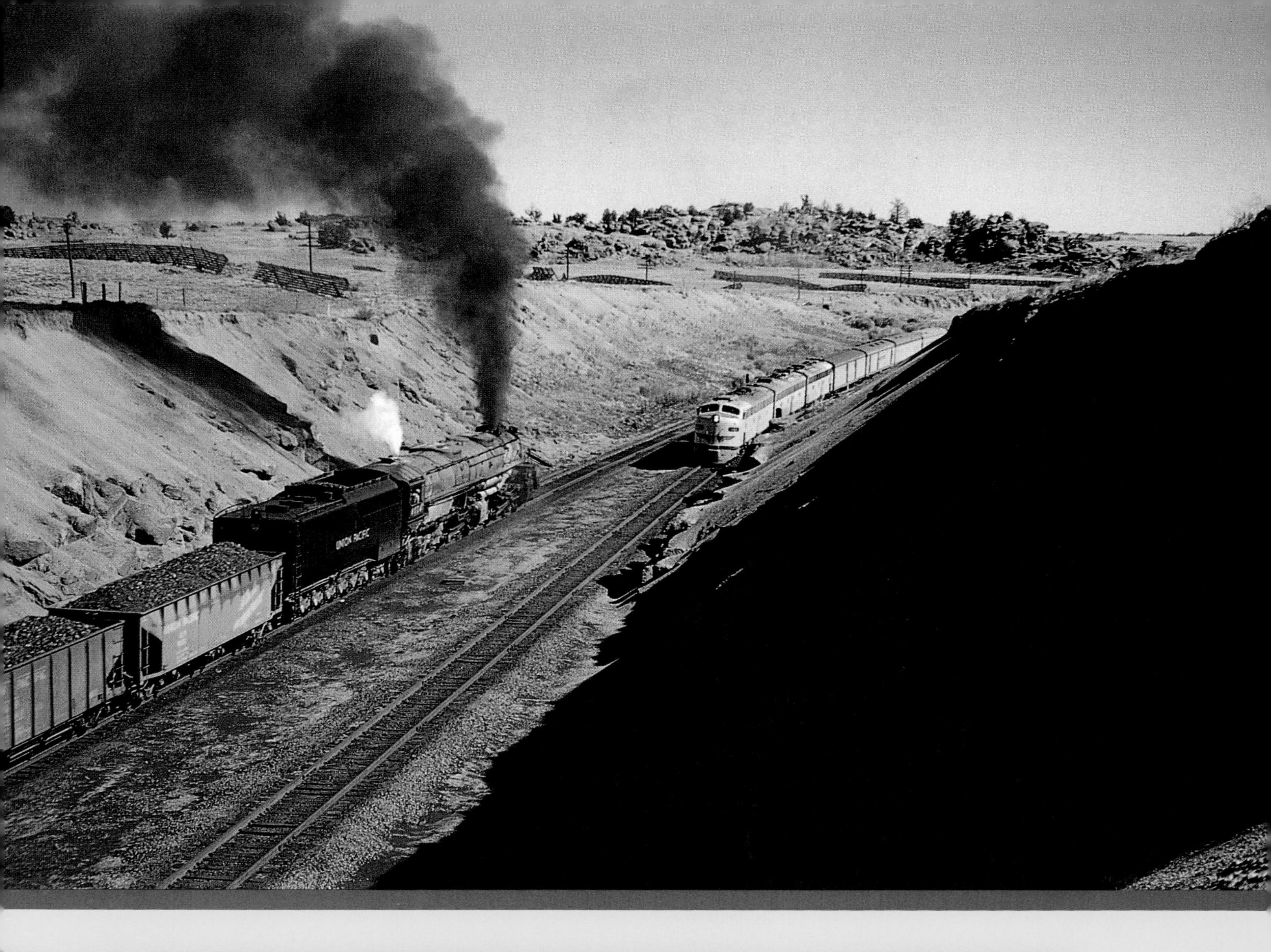
UNION PACIFIC

Union Pacific country

(Opposite page, top) UP engine 4024 was the last Big Boy built - ALCO 1944 - at a cost of $319,600. It powered the eastbound UTX with 95 cars seen previously at Colores. We got it again at the east end of Hermosa Tunnel with left hand operation plus a going-away angle just as the CITY OF SAN FRANCISCO appeared, creating a very pleasing meet shot.
11:25 AM Oct. 2, 1956

(Opposite page, bottom) The afternoon was spent at the west end of Hermosa Tunnel. The unusual power combination of turbine 53 and Big Boy 4006 had the CU/CBF (Clean Up/Council Bluffs Forwarder) with only 56 manifest cars and caboose 3704. 3:28 PM Oct. 2,1956

(Above) The morning of October 4, 1956 dawned crystal clear. We were taking roster shots of 4-6-6-4 #3969 when there was a sudden but pleasant interruption. An eastbound "Green Fruit" had turbine 75 charging out of town with 92 cars (90 of them loaded reefers) and caboose 2701. That evening I headed for home on the CITY OF LOS ANGELES to Chicago and the GTW INTERNATIONAL LIMITED, with an engine cab ride in 4-8-4 number 6406 to Valparaiso, en route to Hamilton and finally the CN/LV MAPLE LEAF to New York.
8:38 AM Oct. 4, 1956

New York Area Potpourri

(Opposite page, top) Snow shots are very nice to get, weather permitting. New York Central's River Division train VW-6 was passing DU tower in Dumont, NJ coming off double-track onto four-track territory. Units 1627-1613 were wheeling 113 cars and caboose 20350. NYC assigned cabooses to specific crews/trains at this time with the 20350 and same crew to return on banana train WB-3. 12:10 PM Feb. 2, 1957

BOSTON • WEEHAWKEN • PHILADELPHIA

READ DOWN BP-2 (1)		TABLE 13		READ UP PB-1 (Daily) (4)
(LS3) 8:45p Mo	Lv.	Boston (B&A)	Ar.	9:40a We
......		Framingham	Ar.	8:45a We
......		Worcester	Ar.	7:45a We
(LS3) 11:59p Mo	Ar.	Springfield	Lv.	5:00a We
(LS3) 1:45a Tu	Lv.	Springfield	Ar.	1:30a We
......		North Adams Jct	Ar.	10:40p Tu
(LS3) 6:30a Tu	Ar.	Selkirk	Lv.	8:15p Tu
(BP2) 11:15a Tu	Lv.	Selkirk	Ar.	5:40p Tu
(BP2) 4:30p Tu (2)	Ar.	Weehawken	Lv.	12:25p Tu (3)
4:30a We	Ar.	Philadelphia (Rdg)	Lv.	7:30p Mo

NOTES

1. LS-3 daily except Sundays . . BP-2 daily except Mondays.
2. Connect to train CNJ/Rdg JB-3.
3. Connection from trains Rdg/CNJ 692.
4. Schedule one hour earlier Springfield and east during periods of Daylight Savings Time.

(Opposite page, bottom) West Shore train PB-1 (Philadelphia to Boston) is shown at one of the most scenic locations in the East. Alco units 1011-3328-1002 had just passed underneath the classic Bear Mountain Bridge and soon ducked into the short Fort Montgomery Tunnel with their even 100 cars and bay window caboose 20345. The Hudson Division is on the opposite side of the majestic river of the same name.
2:28 PM Feb. 2, 1957

(Above) September 6, 1957 was the last time I saw a regular service PRR K4 in action. They were rapidly being phased out by diesels such as these Baldwin sharks 5781A-5772B. Here they have Bay Head Junction to New York Train 706 with 11 cars on the Matawan fill at 7:50 AM. There was a permanent slow order trestle here before the fill was constructed.

BALTIMORE AND OHIO

En Route West

(Opposite page, top) My spring 1957 trip was my first one out to the Rio Grande's narrow gauge. A day en route was spent around Benwood, WV. B&O EM1 number 674 (ex-7624) with an extra west had just crossed the Ohio River into Bellaire, OH heading for Holloway. The 89 cars on the draw-bar required an assist by 2-8-2 #449 with caboose C2257 bringing up the rear.
4:28 PM May 24, 1957

(Opposite page, bottom) B&O engine 755 had an interesting history. It began life as B&M 4108 and was later named *Lily Pons,* honoring the famous diva. It was sold to the B&O, where it got some minor cosmetic surgery, new number 5658 and finally number 755 in the November 1, 1956 system renumbering. Coal train "First 83" from Holloway to Willard was accelerating from the main track coaling facility at Warwick, OH with 60 cars and C2265 totalling 6291 tons.
3:20 PM May 25, 1957

(Above) Nickel Plate Road local freight 21 with Berkshire 756 and 19 cars pulled into the west end of Edgerton, IN siding. Train 35 with the 758 and 67 cars crawled in behind it to clear the main track for eastbound number 36 with the 751 and 100 cars and caboose 1143, creating a rare triple train meet.
3:50 PM May 27,1957

First Contact on the Three-foot Gauge

(Opposite page, top) Stories and photos of the Rio Grande's narrow gauge operation did not excite me but I finally decided to go there. After all, it was an all-steam operation. Engines #486 of class K36 and #495 of class K37 were taking coal at the quaint Chama coal docks. Both 2-8-2's later doubleheaded a freight east to Alamosa.
10:00 AM May 30, 1957

(Above) It wasn't often you could get an overhead view of an engine on a turntable. D&RGW class K28 #476 was on the Durango turntable prior to doubleheading with #478 on a passenger special to Silverton on day two of a three-day fan trip. It was one of my finest rides so I immediately became a narrow-gauge convert.
8:40 AM June 1, 1957

(Opposite page, bottom) The return trip from Durango to Alamosa was only about 30 minutes old when the special showed up at Falfa. The 476, built by ALCO in August, 1923, was working hard with the 11-car train. 8:55 AM June 2, 1957

Fall, 1957

(Above) This photo made the cover of the Winter, 1976 issue of *RAILFAN*, which included an excellent article about the B&O EM1's. Perhaps it deserves repeating. The morning fog was just lifting from the fall foliage at Massillon, OH as an ore extra south with 2-8-8-4 #661 (ex-7611) stepped along with 64 ore cars and caboose C1901.
9:03 AM Oct. 8, 1957

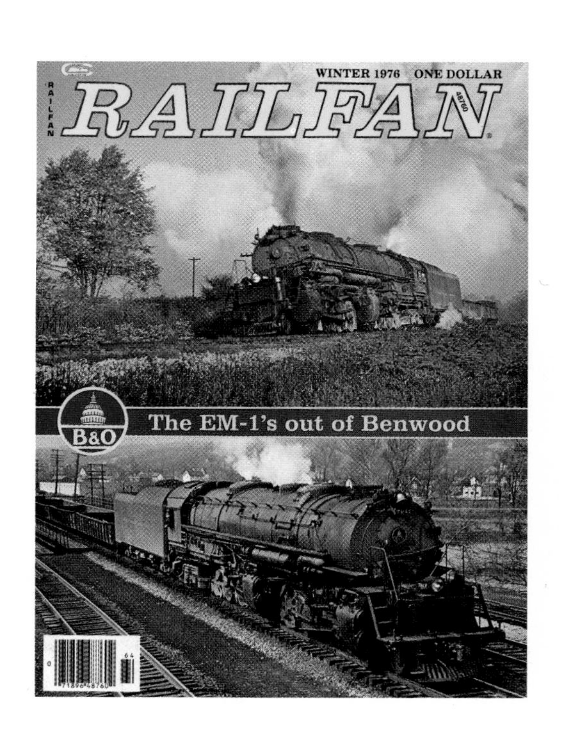

(Opposite page, top) Nickel Plate number 6 was named THE NICKEL PLATE LIMITED, later renamed THE CITY OF CLEVELAND. It was primarily an overnighter with five sleeping cars departing Chicago, with one car going through to Hoboken, NJ on the Lackawanna via Buffalo. ALCO PA's 180-181 had in tow two baggage cars, NKP and DL&W coaches and a DL&W sleeper on the Painesville, OH bridge. 9:27 AM Oct. 9, 1957

(Opposite page, bottom) Did steam power ever handle piggybacks? You bet. Here is proof positive of that fact. It is my version of a *piece de resistance.* NKP second OB-2 was getting out of Conneaut, OH onto the 1320-foot trestle with 2-8-4 #772 and 85 cars including three piggybacks and 28 perishables plus caboose 769. 2:15 PM Oct. 9, 1957

The Route of
Phoebe Snow
Phoebe Snow
Phoebe Snow

N&W in November

(Opposite page, top) My first trip to the Roanoke area, the so-called "Citadel of Steam," was not bad at all. The first train to appear on a crisp morning was an N&W extra east with Y6 2-8-8-2 #2150 helping class A 2-6-6-4 #1236 get 133 cars of coal over the Blue Ridge grade. The cool morning air created wonderful smoke effects at Bonsack, VA.
8:07 AM Nov. 17, 1957

(Opposite page, bottom) It took about four minutes for the rear end to show up with caboose 518300 and Y6 pusher 2158. The heavy-tonnage train made it up the grade in good shape.

(Above) N&W Train 8 was an all-stops local from Roanoke to Norfolk. Streamlined 4-8-2 #128 was assigned to the short consist of a baggage car, an RF&P coach, and three more deadheading baggage cars. It has already topped the Blue Ridge and was now drifting down the east slope by Montvale.
12:15 PM Nov. 17, 1957

NORFOLK AND WESTERN
NORFOLK AND WESTERN
N & W

AMERICAN PRESIDENT LINES LTD

1416

A Winter Day

(Opposite page, top) A sunny day with snow on the ground makes a good combination for train photos. Thursday, January 9, 1958 was such a day. Susquehanna Train 915 was leaving the Erie's Jersey City terminal at 12:45 PM on a leisurely 93-minute run of 37 miles to Butler, NJ. ALCO RS1 #236 was doing the honors with a single Budd-built stainless steel coach. The Susquehanna's main and Erie's Northern Branch trains were the only ones using the almost deserted station at the time.

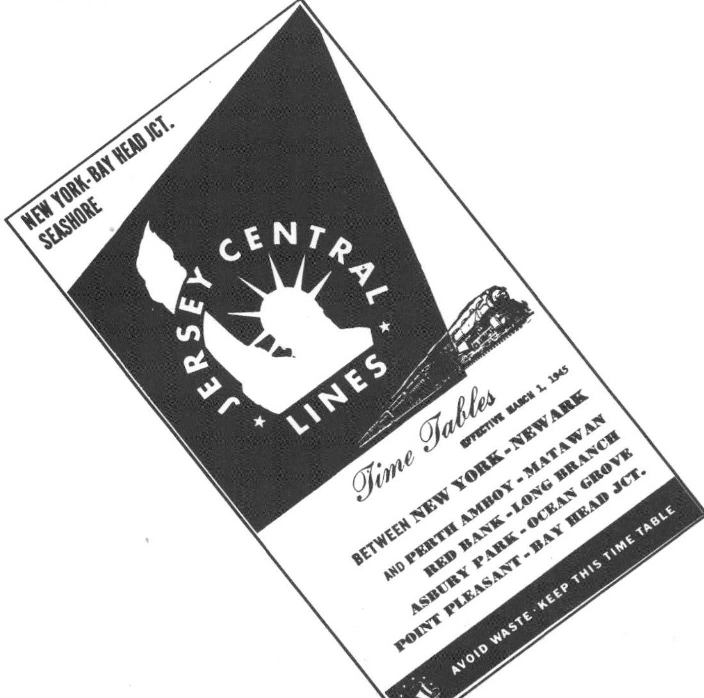

(Above) How do you kill time at the CNJ passenger station while waiting for a B&O train to depart? My solution was to walk out as far as possible on a pier and take a change of pace photo of the ferry *Somerville*. The fantastic downtown New York skyline always made an impressive background. A bonus was an all-white United Fruit "banana boat" in port.

(Opposite page, bottom) B&O number 1, THE NATIONAL LIMITED, was due out of the Jersey City station at 1:55 PM on a 24-hour schedule to St. Louis. It was shown here at Communipaw passing under the Lehigh Valley's National Docks Branch. The classic blue, black and grey 1416 (ex-64A) led a modest 8-car consist.

NORFOLK AND WESTERN

GAINESVILLE MIDLAND

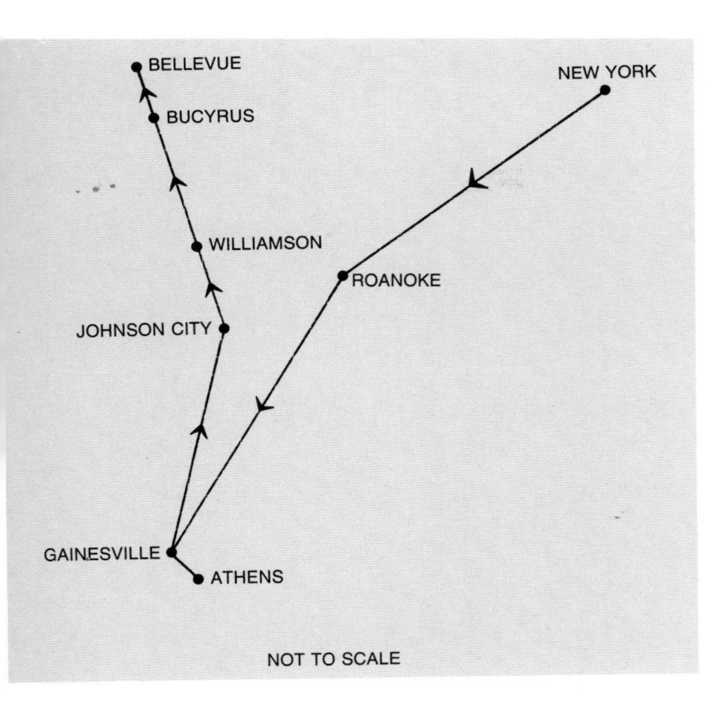

Virginia to Georgia

(Opposite page, top) The first three days of my 1958 trip were spent along the N&W and with good reason - steam was still in service. The N&W's glamour engines were the 14 streamlined 4-8-4's of class J. Train 15 was named THE CAVALIER but it served as a maid-of-all-work from Norfolk to Cincinnati. It passed flag stop Ripplemead, VA with 1941-built 604 and 12 cars.
9:27 AM May 4, 1958

(Below) It was always nice to get a pleasant surprise on a trip. On this occasion, the surprise was that the N&W leased some brand-new PRR GP9's. Freight 85 was exiting the Montgomery, VA tunnel, about 30 miles west of Roanoke, with PRR 7219 on the point ahead of N&W's own year-old GP9's 744 and 758 hauling 90 cars.
12:51 PM May 4, 1958

(Opposite page, bottom) Short lines were not a high priority item with me. However, the Gainesville Midland was so fascinating that three days were spent along their tracks. They operated a day local and night freight on the 41 miles between Gainesville and Athens, GA. The local north with 2-10-0 #206 was creating a mini-volcano lugging only a 10-car train out of Athens.
5:08 PM May 7, 1958

North to Michigan

(Above) The normal coal train consist pulled by a class A 2-6-6-4 westbound from Williamson, WV was 190 cars. That is an extraordinary heavy train on anybody's railroad. N&W's Extra 1225 West was one such "normal" train passing Kermit, WV bound for Portsmouth and Columbus. 10:28 AM May 10, 1958

(Left) While heading north, it was decided to stay overnight in Bucyrus, Ohio if we could find a motel. We almost chose Hotel Elders but eventually found a motel. It was another surprise to see the hotel on fire the next morning and relief that we didn't stay there. 8:00 AM May 11, 1958

(Above) The Grand Trunk Western operated an unheralded commuter train service serving Detroit. The morning run included three "pure" commuter trains plus a standard train from Muskegon and a post-rush hour local from Pontiac. The second commuter was number 72 with 4-6-2 #5634 and ten coaches at Bloomfield Hills, MI. The GTW had the distinction of operating the last steam-powered, regular passenger service in the USA.
8:22 AM May 12, 1958

(Below) GTW freight 512 left Pontiac with 4-8-4 #6333. Thinking it would be nice to get another shot of it, I highballed out of town. A headlight was spotted but the engine wasn't making any smoke. It was this: C&O BL2 #80 on a local freight at Holly. We were on the wrong railroad! 11:30 AM May 12, 1958

1532

3423

Homeward Bound via Canada

(Opposite page, top) Palmerston, Ontario was the CNR hub of several local passenger and mixed trains. Ten wheelers and light Pacifics were the normal engines assigned. A scheduled meet at Bluevale included M331 with 4-6-0 #1530 and thirteen freight cars, a baggage car and coach meeting M332 with sister 1532 and just four freight cars, a baggage car and coach. 1:40 PM May 14, 1958

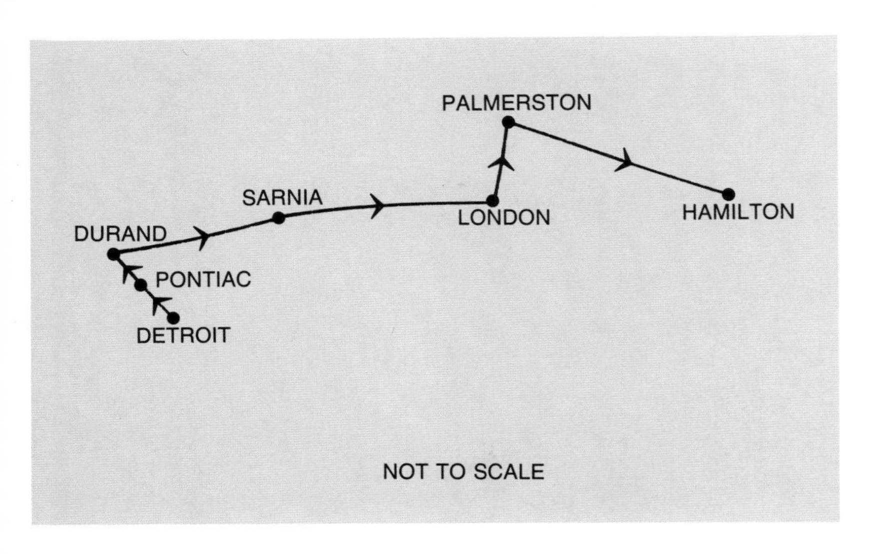

(Opposite page, bottom) The two major styles of streamlining steam engines were the "bullet-nose" and inverted "bathtub." CNR chose the latter. Engine 6400 was an example, powering the six cars of Train 77 on the run from Toronto to London while passing BAYVIEW tower and presently-inactive helper 2-8-2 #3423.
8:54 AM May 17, 1958

(Above) If a picture is worth a thousand words, this is my nomination. What words could possibly describe all the positive points of CNR's handsome engine 6200, the assortment of standard passenger cars and the interlocking serving two railroads. Two towers are visible along the scenic Hamilton Bay with a background of rolling hills topped by clear blue sky as Toronto-to-London Train 83 drifts quietly into Hamilton.
2:40 PM May 16, 1958

1402

all the

Movin' Around

(Opposite page, top) An eastbound Erie freight train had to set off cars at Hillburn, NY effectively blocking the eastbound main track. Number 50 with ALCO PA 859 and five coaches got orders to run east on the westbound main from NEWBURGH Junction to SUFFERN. The operator at SUFFERN tower neglected to hold westbound number 53 with the 1402 and four cars there. The result was this 5:47 AM head-on wreck south of Sloatsburg, NY. Five people were killed and 37 injured. GP7 #1402 was eventually scrapped. 8:55 AM Aug. 11, 1958

(Opposite page, bottom) It was a thrill to get a ride in the cab of Erie E8 #824 on the ERIE LIMITED from Hornell to Port Jervis. During the stop at Binghamton, a Lehigh Valley train for Sayre pulled out of the D&H yard with FT's 506-507, just eight cars and caboose 95006. My publisher says I just "have to" mention those blanked-out portholes!
2:20 PM Jan. 10, 1959

(Above) Meet shots always appealed to me. They were especially satisfying when both trains were moving, as in this instance. CPR number 360 from Detroit with the 1426 and six cars was about to pass CNR number 41 just leaving Toronto for North Bay with 4-8-4 #6238 and seven cars. 9:24 AM Sept. 20, 1958

X-61
UNION PACIFIC

Westward HO

(Opposite page, top) A phone call to Cheyenne revealed that the UP had some Big Boys back in service. Steam was also being used between North Platte and Council Bluffs. That news simply "forced" Bob Collins and I to go out there. We used the Erie's LAKE CITIES to Chicago and the MILW-UP CITY OF LOS ANGELES to North Platte. We captured an eastbound GF (Green Fruit) symbol RV-872 with turbine 61 and GP9's 300B-301 with a high class consist of 3 cattle, 50 perishables and 47 mixed freight plus caboose 2749 leaving town.

1:15 PM Oct. 4, 1958

(Opposite page, bottom) UP number 7 conveyed us to Cheyenne and we rented a car the next morning. There wasn't any UP steam activity but we learned that the C&S had a WHEATLAND TURN out on the road and engine 902 was experiencing oil burner trouble at Chugwater. Well, they got orders to try their luck and that 2-10-2 performed admirably en route south to Cheyenne - seen here at Altus, WY with a short 20-car train. If the 902 was sick, imagine what it could do if all was well!

1:09 PM Oct. 5, 1958

(Above) Here is what we came out for! 4-8-8-4 #4013 was on a drag west with one load, 17 empty freight cars and 105 empty PFE's and caboose 2707. It was performing a major indignity by passing the disabled turbine in Emkay siding, named for contractors Morrison Knudson.

8:27 AM Oct. 7, 1958

Steam's Last Gasp

(Left) The Great Western Railway was a Colorado shortline of about 50 miles between Longmont and Eaton. The commodity moved most was sugar beets. Engine 51, a 2-8-0 built by Baldwin in 1906, did the plant switching at Windsor. Engine 51 survives to this day at the Yakima Valley Rail Museum at Toppenish, WA.

12:50 PM Oct.8, 1958

.....	....	....	....	Mls	*January*, 1953.	....	....	....
.....	....	....	....	0	lve..**Longmont**[1].arr.	....	....	....
Standard—*Mountain time.*	....	....	....	7.9	.... Mead	....	....	....
	....	....	....	17.6	.....**Johnstown**....	....	....	....
				22.5	arr.....**Officer**....lve.			
	Mixed Train Service.			29.0	.....**Loveland**[2].....	**Mixed Train Service.**		
				22.5	lve.....**Officer**....arr.			
				23.3	........**Kelim**........			
	....	....	....	28.0	......**Windsor**[3].....	....	....	....
	....	....	....	33.9	......Severance......	....	....	....
.....	....	....	....	41.8	arr.....**Eaton**[4]...lve.	....	....	....

(Bottom) Big engine 90, a 2-10-0 outshopped by Baldwin in 1924, had the "North End" job of 17 cars and caboose 1007 at Kelim appropriately headed north to end of line at Eaton. The 90 is still in service at Strasburg, PA.

2:10 PM Oct. 8, 1958

(Above) Pacing a steam train is a thrilling adventure. UP 4-8-4 #830, dirty but doing a fine job, looked real good from US Route 30. The 830 had 115 cars in tow heading west a few miles short of North Platte, NE.
2:15 PM Oct. 9, 1958

(Bottom) CLS for "California Live Stock" was a famous UP train running from Council Bluffs to Los Angeles primarily with livestock. There was a rise of 935-feet in the 137.2 miles from Grand Island to North Platte which the CLS was scheduled to do in 145 minutes, or an average of 51.8 MPH. The 829 had a sharp exhaust leaving GI with a car of meat head out, then 28 cars of hogs followed by 43 mixed freight and caboose 3750. About a month later all UP regular service steam operations ceased FOREVER.
11:08 AM Oct. 10, 1958

ERIE
403
808
808

★ ★ ★ LATE

DAILY NEWS

NEW YORK'S PICTURE NEWSPAPER ®

5¢

Vol. 40. No. 71 Copr. 1958 News Syndicate Co. Inc. New York 17, N.Y., Tuesday, September 16, 1958* WEATHER: Mostly fair, warm.

40-60 DEAD IN TRAIN WRECK

(NEWS foto by Sig Stevens)

Norris Fay, a survivor, rests in stretcher on top of pier (arrow) while third passenger coach of train dips into the water. —*Stories pages 3 and 4*

Five Pages of Wreck Fotos—Pages 1, 5, Centerfold and Back Page

Here and There

(Opposite page, top) The O&W was almost a universal favorite among all eastern railfans. The Erie ranked very high also. When the O&W ceased operations in 1957, their FT's were stored in the Erie's Croxton Yard, Secaucus, NJ. O&W FT's 808-808B were built in March 1945. The more senior Erie 403 was still active. The 403 was built by Electro-Motive Corporation as its #953 dated 11-24-39.

3:15 PM Feb.12, 1959

(Opposite page, bottom) The CNJ sadly experienced a sensational and disastrous wreck on September 15, 1958 when a train plunged through an open drawbridge into Newark Bay resulting in 48 deaths. GP7 units 1526 and 1532 powered Train 3314 that fateful day. They were removed from the waters and are shown at the Communipaw, Jersey City engine terminal.

12:50 PM Feb. 6, 1959

(Below) The Burlington ran an excellent fan trip from Chicago to Galesburg and return with 4-8-4 #5632 on April 5, 1959. A stop at the Galesburg engine terminal was featured with an assortment of FT and F3 units.

622

4
LEHIGH AND HUDSON RIVER

Maybrook Trains

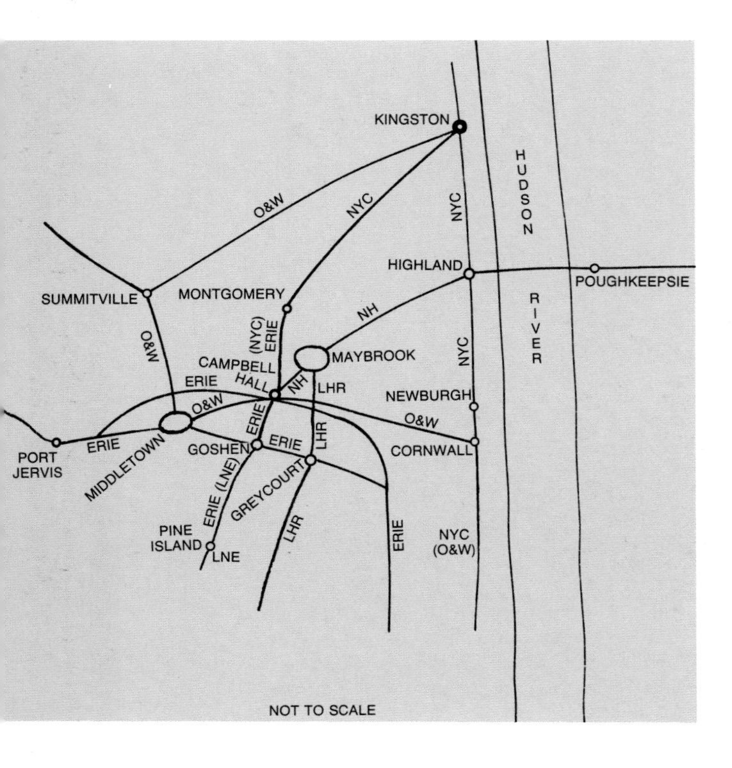

(Opposite page, top) The small towns of Maybrook and Campbell Hall, NY were once acclaimed as a railfan wonderland. With good reason. After all, six major railroads congregated there including: New Haven, Erie, Lehigh & Hudson River, Lehigh & New England, New York, Ontario & Western, and New York Central. NYO&W freight BC-1, with a connection from Boston off NH train BO-1, was leaving Maybrook behind F3 #822 and FT's 803B-803, only 27 cars and caboose 8342.
11:56 AM Dec. 11, 1953

(Opposite page, bottom) Symbol train HO-6 started at Hagerstown,MD and traveled on the rails of five railroads WM to Shippensburg, PA, RDG to Allentown, CNJ to Easton, PRR to Belvidere, NJ, and finally L&HR to Maybrook. L&HR HO-6 was approaching Maybrook Yard behind ALCO RS3's 4 and 9. HO-6 this trip started with 50 cars out of Allentown and picked up 64 more at Phillipsburg, NJ off the Lehigh Valley. Caboose 14 carried the markers.
12:15 PM Feb. 23, 1958

(Below) The Lehigh & New England did not operate any symbol trains - just extras. However, they ran a "day job" between Pen Argyl, PA and Maybrook that was called the "local." A night train called "the cement train" also was run. Here is the "local" heading back to Pen Argyl behind a trio of ALCO FA/FB's 707-752-704 with 36 cars and the 573 passing near Orange Farm, NY. The L&NE terminated all operations on March 31, 1961.
5:07 PM March 7, 1959

CNR

Canadian National

(Opposite page, top) The Canadian National's first diesel paint scheme was an attractive green, yellow and black with gold striping. ALCO units 9414A-9415B had a short westbound 18-car train coming into St. Lambert, Que.
8:53 AM Sept. 16, 1955

(Opposite page, bottom) CNR Train 75 was named THE FOREST CITY. Its schedule called for a 3:30 PM departure from Toronto and 5:40 PM arrival in London, making it the fastest train on that route account avoiding Hamilton. The CNR acquired twenty 4-8-2's from Montreal Locomotive Works in 1944, their last new steam power. Each were nicknamed "Bullet Nose Betty." Those nose cones were removed from engines operating in the Prairie Region. The 6076 had full control of the six cars going by Hamilton West.
4:25 PM Aug. 21. 1957

(Above) The CNR had only five 4-6-4's. With 80-inch drivers, they contrasted with the CPR which owned 65 of that wheel arrangement but with 75-inch drivers. Engines 5700-5704 were originally put in service on the fast Montreal-Toronto trains until displaced by the streamlined 4-8-4's. Most 5700's had a distinctive capped stack, disc drivers and a rounded tender. The 5700 had the six cars of number 77 smoking up the landscape by BAYVIEW interlocking. 9:09 AM March 20, 1959

Canadian Pacific

(Above) The DOMINION was the Canadian Pacific's premier transcontinental passenger train until the 1955 inauguration of the CANADIAN, when it became the secondary train. Number 8 was nearing its Montreal terminal led by FP7/F7B 1416-1910 and eight cars including two domes. 9:10 AM March 28, 1958

(Below) Forty-five of the 65 CPR 4-6-4's were semi-streamlined in a unique design with an abundant application of burgundy paint. Class HIC 2828, built in October, 1937 by Montreal Locomotive Works, awaited assignment to Train 235 in Montreal's Glen Yard. 11:20 AM March 28, 1959

(Above) The CPR's three 4-4-0's operating in New Brunswick acquired legendary status as the oldest operating steam locomotives at that time. Numbers 29, 136 and 144 took turns powering a mixed train between Norton and Chipman. The 136 had the assignment this day at Cody with three freight cars and a combine. The 136 was built by Rogers in 1883 and rebuilt by CPR in 1914 as class A2m. 10:19 AM July 21, 1958

(Below) Hamilton is located at a low level along Lake Ontario and is almost completely surrounded by hilly obstacles. CPR trains encountered a stiff grade en route north to GUELPH Jct. The wayfreight's 2-8-2 #5406 emitted an awesome smoke display with only a 14-car train. 2:30 PM April 4, 1959

PAPERS
SUPPLYING PAPER PRODUCTS

1959 Trip

(Opposite page, top) I rode a PRR overnight train from New York to Washington to start my 1959 vacation trip. B&O's number 11, THE METROPOLITAN SPECIAL, was my transportation to Cincinnati where I took the NATIONAL LIMITED the following morning to St. Louis. During the stopover there, I got a rear end shot of the Wabash BANNER BLUE leaving the city with the 1021-1003 and a modest five-car consist with parlor open-end observation car *City of Wabash* displaying the markers. The MOP's COLORADO EAGLE was my next conveyance.
2:50 PM May 27, 1959

(Opposite page, bottom) Once you get a taste of the Rio Grande's narrow gauge steam operation it is hard to resist making repeat visits. 1959 was the year of my third consecutive trip there. A "Cumbres Turn" was caught grinding up the mountain at Coxo, CO with the 498 on the point of just 18 loads with pusher 483 and caboose 0503.
2:25 PM May 29, 1959

(Above) The ride between Durango and Silverton ranks as one of the very best in the country, even to this day. A doubleheader of 476 and 473 had the 12-car special moving right along going by the Hermosa water tank.
8:45 AM May 30, 1959

DAUPHIN
OCHRE RIVER
PORTAGE LA PRAIRIE
RIVERS
WINNIPEG
CP
GN
ST. PAUL
CNW
OMAHA
UP
DENVER
PUEBLO
MOP
KANSAS CITY
MOP
B&O
ST. LOUIS
LVERTON
RANGO
ALAMOSA
NOT TO SCALE

Iowa Manitoba Maine

(Below) The UP's CITY OF PORTLAND was taken from Denver to Omaha on June 1. The following day I was aboard C&NW number 204, THE NORTH AMERICAN, with E7A 5019B and nine cars bound for St. Paul. The train took siding at Whiting, IA for freight 17 with F7's 6503C-4070A and 61 cars.
11:57 AM June 2, 1959

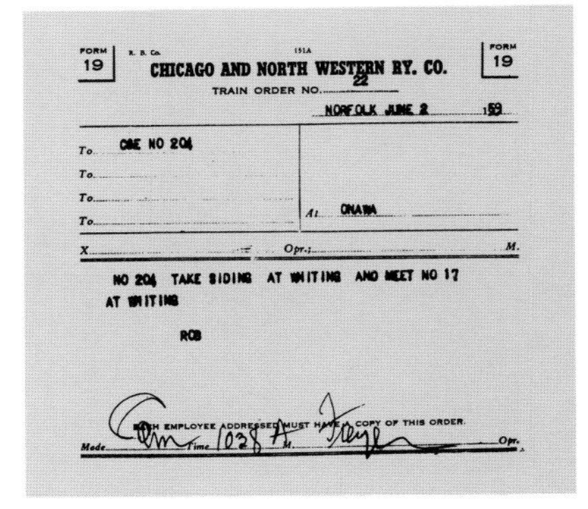

FORM 19 R. B. Co. 151A FORM 19

CHICAGO AND NORTH WESTERN RY. CO.

TRAIN ORDER NO. 22

NORFOLK JUNE 2 1959

To C&E NO 204

To

To

To At ONAWA

X Opr. M.

NO 204 TAKE SIDING AT WHITING AND MEET NO 17
AT WHITING

RCB

EACH EMPLOYEE ADDRESSED MUST HAVE A COPY OF THIS ORDER.

Made Com Time 1038 A M. Opr.

(Opposite page, top) GN number 7 was my next ride north to Winnipeg in Manitoba. A friend met me there and we did some serious train chasing for several days. One of my favorite shots of the trip was this of CNR freight 412 from Dauphin to Winnipeg. It was comparatively rare to see a Pacific hauling freight - number 5620 in this case with 44 cars passing the distinctive wooden water tank at Ochre River.
11:57 AM June 6, 1959

(Opposite page, bottom) A whirlwind trip to McAdam in New Brunswick was marred by unfavorable weather. The conditions improved on the way back home. A quick stop in Waterville, ME resulted in a Maine Central westbound way freight with GP7 575 going by the static-displayed 4-6-2 number 470.
2:35 PM Aug. 1, 1959

5632

2541

Fan Trips

(Opposite page, top) It is spelled like Altoona, PA but the action is at Altona, IL. The big attraction of an Illini Railroad Club sponsored fan trip on the Burlington between Chicago and Galesburg was steam locomotive 5632. This 4-8-4 oil-burner was in class 05B and made a photo run for the benefit of both photographers and riders.
1:29 PM April 5, 1959

(Opposite page, bottom) There was a fan trip on the CNR from mainline Belleville to branchline Bancroft with steam engines 90 and 2649 on May 10, 1959. The previous day the CPR way freight to Farnham was stopped at Delson, Quebec just as D&H's MONTREAL LIMITED was going by with the 4023-4009 and 11 cars including six Pullmans.
7:40 AM May 9, 1959

(Above) The Reading ran one of its famous *Rail Rambles* from WAYNE Junction in Philadelphia to Shamokin on October 25, 1959. The timing was perfect for the fall foliage. The class T1 #2124 had the16-car consist rolling right along just a few miles short of Tamaqua.

Their Days are Numbered

(Below) Most Western movies ended with the good-guy riding off into the sunset and all was well with the world. Steam locomotives during the 1950's also rode off into the sunset - right to the scrap line. Here, C&O 2-6-6-6 #1644 is heading west into the setting sun on one of her last trips - suffering from lack of maintenance but still highly functional. C&O ended all steam operations several months later - in September 1956. Sister engine #1601 is preserved at the Henry Ford Museum in Dearborn, MI. 5:20 PM June 6, 1956

(Opposite page, top) A freight train does not look like a train without a caboose on the rear end. Most people are deeply saddened that the railroads are eliminating them. The rear end of a Reading extra west has the traditional look of a train struggling towards Locust Summit, PA. The three F7's headed by the 269 had a consist of 50 cars - mostly coal - and caboose 92901, too heavy to make it alone up the 2.6% grade from Gordon, so class T1 4-8-4 #2113 was assigned to assist. The 2113 was one of nine engines leased shortly later to the PRR. Several T1's have been preserved including the 2102 still seeing service in 1991 on the Reading and Northern. 12:41 PM May 9, 1956

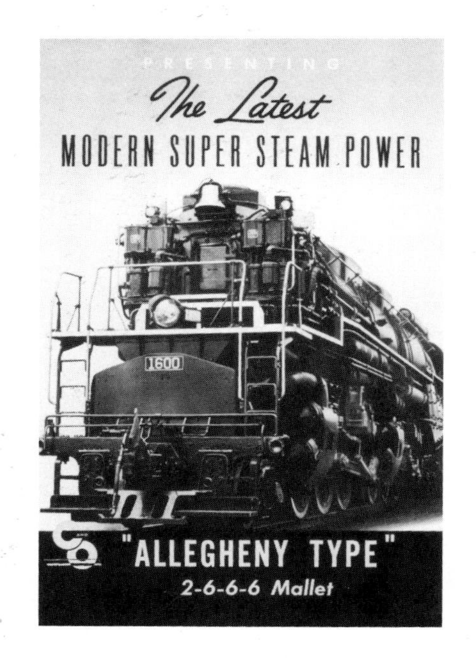

(Opposite page, bottom) This photo of GTW's "Pontiac Turn" returning to Durand depicts some lost arts. The operator is handing up orders to the rear end crew in a wooden-sheathed caboose. Perhaps the "op" got his instructions from the dispatcher on telegraph - but more likely verbally. The steam locomotive would soon be gone along with that wooden caboose. Train orders were becoming out-moded account of more modern practices and operators were soon put on the endangered species list due to more sophisticated hardware. What a way to end a decade! 8:00 AM May 13, 1958